HOW TO
FIX
THE
FUTURE

HOW TO
FIX
THE
FUTURE

ANDREW KEEN

Atlantic Monthly Press
New York

FIRST EDITION

Published simultaneously in Canada
Printed in the United States of America

First Grove Atlantic hardcover edition: February 2018

This book was set in 11.75 pt. Jenson Text LT
by Alpha Design & Composition of Pittsfield, NH.

Library of Congress Cataloging-in-Publication data is available for this title.

ISBN 978-0-8021-2664-1
eISBN 978-0-8021-8912-7

Atlantic Monthly Press
an imprint of Grove Atlantic
154 West 14th Street
New York, NY 10011

Distributed by Publishers Group West

groveatlantic.com

18 19 20 21   10 9 8 7 6 5 4 3 2 1

For our kids

CONTENTS

"The Senate also has a standing rule never to debate a matter on the same day that it is first introduced but to put it off till the next morning. This they do so that a man will not blurt out the first thought that occurs to him, and then devote all his energies to defending his own proposals, instead of considering the common interest. They know that some men have such a perverse and preposterous sense of shame that they would rather jeopardize the general welfare than their own reputation by admitting they were short-sighted in the first place. They should have had enough foresight at the beginning to speak with consideration rather than haste."

—*Thomas More*, Utopia[1]

PREFACE

AN INTERNET OF PEOPLE

Having spent the last decade writing critically about the digital revolution, I've been called everything from a Luddite and a curmudgeon to the "Antichrist of Silicon Valley." At first I was part of a small group of dissenting authors who challenged the conventional wisdom about the internet's beneficial impact on society. But over the last few years, as the zeitgeist has zigged from optimism to pessimism about our technological future, more and more pundits have joined our ranks. Now everyone, it seems, is penning polemics against surveillance capitalism, big data monopolists, the ignorance of the online crowd, juvenile Silicon Valley billionaires, fake news, antisocial social networks, mass technological unemployment, digital addiction, and the existential risk of smart algorithms. The world has caught up with my arguments. Nobody calls me the Antichrist anymore.

Timing—as I know all too well from my day job as a serial entrepreneur of mostly ill-timed start-ups—is everything. Having written three books exposing the dark side of the digital revolution, I think the time is now right for something more positive. So, rather than yet another noxious screed against contemporary technology, this book offers what I hope are constructive answers to the myriad questions on the digital horizon. To borrow a fashionable Silicon Valley word, this represents a pivot in my writing career. What you are about to read is a solutions book. It's obvious that the future needs to be fixed. The question now is *how* to fix it.

This is also a people book. What I've tried to write is a human narrative. It's the story of how people all over the world—from Estonia and Singapore to India, Western Europe, the United States, and beyond—are trying to solve the great challenges of our digital age. "Out of the crooked timber of humanity," the eighteenth-century German philosopher Immanuel Kant suggested, "no straight thing was ever made." But there is one *straight thing* about all the people described in this book. Although there might not be any single solution, any magic bullet for creating an ideal network society, what unites all these people is their determination—what I call "agency"—to shape their own fate in the face of technological forces that often seem both uncontrollable and unaccountable.

There is today much hype, some of it justified, about the "internet of things"—the network of smart objects that is the newest new thing in Silicon Valley. Rather than an internet of things, however, this book showcases an internet of people. I show that instead of smart technology, it's smart human beings, acting as they've always done throughout history—as innovators, regulators, educators, consumers, and, above all, as engaged citizens—who are fixing the twenty-first-century future. At a time when our traditional notion of "humanity" is threatened by artificial intelligence (AI) and other smart technologies, it's this humanist verity—the oldest of old things—that is the central message of the book.

There is, however, nothing inevitable about a global network of people successfully fixing the future. The issues that confront us are urgent and complex. Time is many things, but it isn't infinite, at least not for us humans. The digital clock, which seems to proceed at a more accelerated pace than its analog forebear, is already ticking furiously. Unless we act now, we increasingly risk becoming powerless appendages to the new products and platforms of Big Tech corporations. This book, then, is also a call to arms in a culture infected by a creeping (and creepy) technological determinism. And it's a reminder that our own human agency—our timeless responsibility to shape our own societies—is essential if we are to build a habitable digital future.

In contrast with smart cars, the future will never be able to drive itself. None of us, not even the Antichrist of Silicon Valley, have superhuman powers. But by working together, as we've done throughout history, we can build a better world for our children. This book is dedicated to them. They are why the future matters.

Andrew Keen
Berkeley, California
July 2017

HOW TO
FIX
THE
FUTURE

INTRODUCTION

WE'VE BEEN HERE BEFORE

The future, it seems, is broken. We are caught between the operating systems of two quite different civilizations. Our old twentieth-century system doesn't work anymore, but its replacement, a supposedly upgraded twenty-first-century version, isn't functioning properly either. The signs of this predicament are all around us: the withering of the industrial economy, a deepening inequality between rich and poor, persistent unemployment, a fin-de-siècle cultural malaise, the unraveling of post–Cold War international alliances, the decline of mainstream media, the dwindling of trust in traditional institutions, the redundancy of traditional political ideologies, an epistemological crisis about what constitutes "truth," and a raging populist revolt against the establishment. And while we are all too familiar with what is broken, we don't seem to know how we can get anything to work anymore.

What is causing this great fragmentation? Some say too much globalization, others say not enough. Some blame Wall Street and what they call the "neoliberalism" of free

1

market monetary capitalism, with its rapacious appetite for financial profit. Then there are those who see the problem in our new, unstable international system—for instance, the cult-of-personality authoritarianism in Russia, which they say is destabilizing Europe and America with a constant barrage of fake news. There's the xenophobic reality television populism of Donald Trump and the success of the Brexit plebiscite in the United Kingdom—although sometimes it's hard to tell if these are causes or effects of our predicament. What is clear, however, is that our twentieth-century elites have lost touch with twenty-first-century popular sentiment. This crisis of our elites explains not only the scarcity of trust bedeviling most advanced democracies but also the populist *ressentiment* on both left and right, against the traditional ruling class. Yet it also feels as if we are all losing touch with something more essential than just the twentieth-century establishment. Losing touch with ourselves, perhaps. And with what it means to be human in an age of bewilderingly fast change.

As Steve Jobs used to say, teasing his audience before unveiling one of Apple's magical new products, there's "one more thing" to talk about here. And it's the biggest *thing* of all in our contemporary world. It is the digital revolution, the global hyperconnectivity powered by the internet, that lies behind much of the disruption.

In 2016, I participated in a two-day World Economic Forum (WEF) workshop in New York City about the "digital transformation" of the world. The event's focus was on what it called the "combinatorial effects" of all these new internet-based technologies—including mobile, cloud, artificial intelligence, sensors, and big data analytics. "Just as the steam engine and electrification revolutionized entire sectors of the economy from the eighteenth century onward," the seminar concluded, "modern technologies are beginning to dramatically alter today's industries."[1] The economic stakes in this great transformation are dizzying. Up to $100 trillion can be realized in the global economy by 2025 if we get the digital revolution right, the WEF workshop promised.

And it's not only industry that is being *dramatically* changed by these digital technologies. Just as the industrial revolution transformed society, culture, politics, and individual consciousness, so the digital revolution is changing much about twenty-first-century life. What's at stake here is worth considerably more than just $100 trillion. Today's structural unemployment, inequality, anomie, mistrust, and the populist rage of our anxious times are all, in one way or another, a consequence of this increasingly frenetic upheaval. Networked technology—enabled in part by Jobs's greatest invention, the iPhone—in combination with other digital technologies and devices, is radically disrupting our political, economic, and social lives. Entire industries—education, transportation, media, finance, health care, and hospitality—are being turned upside down by this digital revolution. Much of what we took for granted about industrial civilization—the nature of work, our individual rights, the legitimacy of our elites, even what it means to be human—is being questioned in this new age of disruption. Meanwhile, Silicon Valley is becoming the West Coast version of Wall Street, with its multibillionaire entrepreneurs taking the role of the new masters of the universe. In 2016, for example, tech firms gave out more stock-based compensation than Wall Street paid out in bonuses.[2] So, yes, our new century is turning out to be the networked century. But, to date, at least, it's a time of ever-deepening economic inequality, job insecurity, cultural confusion, political chaos, and existential fear.

We've been here before, of course. As the "digital transformation" WEF workshop reminds us, a couple of hundred years ago the similarly disruptive technology of the industrial revolution turned the world upside down, radically reinventing societies, cultures, economies, and political systems. The nineteenth-century response to this great transformation was either a *yes*, a *no*, or a *maybe* to all this bewildering change.

Reactionaries, mostly Luddites and romantic conservatives, wanted to destroy this new technological world and return to what appeared to them, at least, to be a more halcyon era. Idealists—including, ironically enough, both uncompromisingly free market capitalists and revolutionary communists—believed that the industrial technology would, if left to unfold according to its own logic, eventually create a utopian economy of infinite abundancy. And then there were the reformers and the realists—a broad combination of society, including responsible politicians on both the left and the right, businesspeople, workers, philanthropists, civil servants, trade unionists, and ordinary citizens—who focused on using human agency to fix the many problems created by this new technology.

Today we can see similar responses of *yes*, *no*, or *maybe* to the question of whether the dramatic change swirling all around us is to our benefit. Romantics and xenophobes reject this globalizing technology as somehow offending the laws of nature, even of "humanity" itself (an overused and under-defined word in our digital age). Both Silicon Valley techno-utopians and some critics of neoliberalism insist that the digital revolution will, once and for all, solve all of society's perennial problems and create a cornucopian postcapitalist future. For them, much of this change is inevitable—"*The Inevitable*"³ according to one particularly evangelical determinist. And then there are the *maybes*, like myself—realists and reformers rather than utopians or dystopians—who recognize that today's great challenge is to try to fix the problems of our great transformation without either demonizing or lionizing technology.

This is a *maybe* book, based on the belief that the digital revolution can, like the industrial revolution, be mostly successfully tamed, managed, and reformed. It hopes that the best features of this transformation—increased innovation, transparency, creativity, even a dose of healthy disruption—might make the world a better place. And it outlines a series of legislative, economic, regulatory, educational, and ethical reforms that can, if implemented correctly, help fix our common future. Just as the digital revolution is being driven by what that WEF workshop called the "combinatorial effects" of several networked technologies, solving its many problems requires an equally combinatorial response. As I've already argued, there is no magic bullet that can or will ever create the perfect society—digital or otherwise. So relying on a single overriding solution—a perfectly free market, for example, or ubiquitous government regulation—simply won't work. What's needed, instead, is a strategy combining regulation, civic responsibility, worker and consumer choice, competitive innovation, and educational solutions. It was this multifaceted approach that eventually fixed many of the most salient problems of the industrial revolution. And today we need an equally combinatorial strategy if we are to confront the many social, economic, political, and existential challenges triggered by the digital revolution.

Maybe we can save ourselves. Maybe we can better ourselves. But only maybe. My purpose in this book is to draw a map that will help us find our way around the unfamiliar terrain of our networked society. I traveled several hundred thousand miles to research that map—flying from my home in Northern California to such faraway places as Estonia, India, Singapore, and Russia, as well as to several Western European countries and many American cities outside California. And I interviewed close to a hundred people in the many places I visited—including presidents, government ministers, CEOs of tech start-ups, heads of major media companies, top antitrust and labor lawyers, European Union commissioners, leading venture capitalists, and some of the most prescient futurists in the world today. The wisdom in this book is theirs. My role is simply to join the dots in the drawing of a map that they have created with their actions and ideas.

One of the most prescient people at the 2016 WEF workshop was Mark Curtis, a serial start-up entrepreneur, writer, and design guru who is also cofounder of Fjord, a London-based creative agency owned by the global consultancy firm Accenture. "We need an optimistic map of the future which puts humans in its center," Curtis said to me when I later visited him at the Fjord office near Oxford Circus in London's West End. It's a map, he explained, that should provide guidance for all of us about the future—establishing in our minds the outlines of an unfamiliar place so that we can navigate our way around this new terrain.

This book, I hope, is that map. From old carpet factories in Berlin to gentlemen's colonial clubs in Bangalore to lawyers' offices in Boston to the European Commission headquarters in Brussels and beyond, *How to Fix the Future* offers a new geography of how regulators, innovators, educators, consumers, and citizens are fixing the future. But there's no Uber or Lyft–style service that can whisk us, with the click of a mouse or the swipe of a finger, into the future. No, not even the smartest technology can solve technological problems. Only people can. And that's what this book is about. It is the story of how some people in some places are solving the thorniest problems of the digital age. And how their example can inspire the rest of us to do so too.

Chapter One

More's Law

Agency

This nineteenth-century room is full of twenty-first-century things. The room itself—the entire top floor of what was once a Berlin factory—is decrepit, its brick walls shorn of paint, its wooden floors splintered, the pillars holding up its low ceiling chipped and cracked. The four-story brick building, one of Berlin's few remaining nineteenth-century industrial monuments, is named the Alte Teppichfabrik (the Old Carpet Factory). But, like so much else of old Berlin, this industrial shell is now filled with new people and new technology. This crowd of investors, entrepreneurs, and technologists are all staring at a large electronic screen in front of them. It is broadcasting the image of a bespectacled young man with a pale, unshaven face, staring intently into a camera. Everyone in the room is watching him speak. They are all listening raptly to the most notorious person in cyberspace.

"What we are losing is a sense of agency in our societies," he tells them. "That's the existential threat we all face."

The whole spectacle—the dilapidated room, the mesmerized audience, the pixelated face flickering on the giant screen—recalls for me one of television's most iconic commercials, the Super Bowl XVIII slot for the Apple Macintosh computer. In this January 1984 advertisement for the machine that launched the personal computer age, a man on a similarly large screen in a similarly decrepit room addresses a crowd of similarly transfixed people. But in the Macintosh commercial the man is a version of Big Brother, the omniscient tyrant from Orwell's twentieth-century dystopian novel *Nineteen Eighty-Four*. The young man on the Berlin screen, in contrast, is an enemy of authoritarianism. He is someone who, at least in his own mind, is a victim rather than a perpetrator of tyranny.

His name is Edward Snowden. A hero to some and a traitorous hacker to others, he is the former CIA contractor who, having leaked classified information about a series of US government surveillance programs, fled to Vladimir Putin's Russia and now mostly communicates with the outside world through cyberspace.

The Berlin audience has come to the old carpet factory for a tech event titled "Encrypted and Decentralized," organized by the local venture firm BlueYard Capital. Its purpose—like that of this book—is to figure out how to fix the future. "We need to encode our values not just in writing but in the code and structure of the internet," the invitation to the event had said. Its goal is to insert our morality into digital technology so that the internet reflects our values.

Snowden's electronic face on the Berlin screen is certainly a portrait of human defiance. Staring directly at his

German audience, he repeats himself. But this time, rather than an observation about our collective powerlessness, his message is more like a call to arms.

"Yes, what we are losing," he confirms, "is a sense of agency in our society."

It is perhaps appropriate that he should be offering these thoughts from cyberspace. The word "cyberspace" was coined by the science fiction writer William Gibson in his 1984 novel *Neuromancer* and was invented to describe a new realm of communication among personal computers such as the Apple Macintosh. Gibson adapted it from the word "cybernetics," a science of networked communications invented by the mid-twentieth-century Massachusetts Institute of Technology (MIT) mathematician Norbert Wiener. And Wiener named his new science of connectivity after the ancient Greek word *kybernetes*, meaning a steersman or a pilot. It was no coincidence that Wiener—who, along with fellow MIT alumni Vannevar Bush and J.C.R. Licklider,[1] is considered a father of the internet—chose to name his new science after *kybernetes*. Networked technology, Wiener initially believed, could *steer* or *pilot* us to a better world. This assumption, which Wiener shared not only with Bush and Licklider, but with many other twentieth-century visionaries—including Steve Jobs and Steve Wozniak, the cofounders of Apple—was based on the conviction that this new technology would empower us with agency to change our societies. *"You'll see why 1984 won't be like '1984,'"* promised the iconic Super Bowl XVIII advertisement about the transformative power of Jobs's and Wozniak's new desktop computer.

But Edward Snowden's virtual speech at the Alte Tep-pichfabrik doesn't share this optimism. Communicating in cyberspace, presumably from a Russian safe house a couple of thousand miles east of the German capital, Snowden is warn-ing his Berlin audience that contemporary technology—the power of the network, in an age of ubiquitous computing, to snoop on and control everything we do—is undermining our power to govern our own society. Rather than a steers-man, it has become a jailor.

"Individual privacy is the right to the self. It's about power. It's about the need to protect our reputation and be left alone," Snowden tells the Berlin audience from cyber-space. In this nineteenth-century room, he is articulating a classic nineteenth-century sensibility about the inviolability of the self.

From somewhere in Putin's Russia, Edward Snowden poses a question to his Berlin audience to which he knows the answer. "What does it mean," he asks, "when we are all transparent and have no secrets anymore?"

In Snowden's mind, at least, it means that we don't exist anymore. Not in the way that nineteenth-century figures, like William Wordsworth or Henry James, regarded our intrinsic right to privacy.[2] It's the same argument that two American lawyers, Samuel Warren and Louis Brandeis, made in their now iconic 1890 *Harvard Law Review* article, "The Right to Privacy." Written as a reaction to the then radi-cally disruptive new technology of photography, the Boston-based Warren and Brandeis (who would later become a US Supreme Court justice) argued that "solitude and privacy have been more essential to the individual." The right to "be let alone," they thus wrote, was a "general right to the immunity of the person . . . The right to one's personality."[3]

So how do we restore nineteenth-century values to twenty-first-century life? How can agency be reinvented in the digital age?

At the climax of that 1984 advertisement for the Macintosh, a vigorous blonde in red-and-white workout gear bursts into the decrepit room and, hurling a hammer at the screen, blows up the image of Big Brother. She isn't a Luddite, of course; the whole point of this one-minute Super Bowl slot, after all, was to convince its millions of viewers to spend $2,500 on a new personal computer. But the Apple commercial does remind us, albeit through Madison Avenue's Technicolor-tinted lenses, about the central role of human agency in changing the world and in keeping us safe from those who would take away our rights.

The issue the virtual Edward Snowden is raising with his Berlin audience is also the central question in this book. How can we reassert our agency over technology? How do we become like that vigorous blonde in the Macintosh advertisement and once again make ourselves the pilots of our own affairs?

Moore's Law

Edward Snowden is right. The future isn't working. There's a hole in it. Over the last fifty years we've invented transformational new technologies—including the personal computer, the internet, the World Wide Web, artificial intelligence, and virtual reality—that are transforming our society. But there is one thing that's missing from this data-rich world. One thing that's been omitted from the new operating system.

Ourselves. We are forgetting about our place, the *human* place, in this twenty-first-century networked world. That's where the hole is. And the future, *our* future, won't be fixed until we fill it.

Everything is getting perpetually upgraded except us. The problem is there's no human version of Moore's Law, the 1965 prediction by Intel cofounder Gordon Moore that the processing power of silicon chips would double about every eighteen months.[4] Today, half a century after Gordon Moore described the phenomenon that would later be named Moore's Law,[5] it remains the engine driving what the Pulitzer Prize–winning author Thomas Friedman calls our "age of acceleration."[6] So, yes, that iPhone in your pocket may be unrecognizably faster, and more connected, powerful, and intelligent, than its predecessor, the once-revolutionary Apple Macintosh personal computer, let alone a mid-sixties multimillion-dollar mainframe machine that required its own air-conditioned room to operate. But in spite of promises about the imminent merging of man and computer by prophets of the "Singularity"—such as Google's chief futurist, Ray Kurzweil, who still insists that this synthesis will inevitably happen by 2029—we humans, for the moment at least, are no speedier, no smarter, and, really, no more self-aware than we were back in 1965.

What Friedman euphemistically dubs a "mismatch" between technology and humanity is, he says, "at the center of much of the turmoil roiling politics and society in both developed and developing countries today . . . [and]

now constitutes probably the most important governance challenge across the globe."[7] As Joi Ito, the director of the MIT Media Lab, warns, when everything is moving quickly except us, the consequence is a social, cultural, and economic "whiplash."[8]

Few people have given this asymmetry more thought than the philosopher whom Thomas Friedman acknowledges as his "teacher" in these matters, Dov Seidman, author of *How* and the CEO of LRN, which advises companies on ethical behavior, culture, and leadership.[9, 10]

Seidman reminds us that "there's no Moore's Law for human progress" and that "technology can't solve moral problems." Most of all, however, he has taught me in our numerous conversations that the hyperconnected twenty-first-century world hasn't just changed, but has been totally reshaped. And since this reshaping has occurred faster than we have reshaped ourselves, Seidman says, we now need to play "moral catch-up."

Seidman describes a computer as a "brain outside of ourselves," our "second brain." But, he warns, from an evolutionary standpoint, there's been what he calls an "exponential leap," and this new brain has outpaced our heart, our morality, and our beliefs. We have become so preoccupied looking down at our second brains, he warns, that we are forgetting how to look smartly at ourselves. As these devices get faster and faster, we appear to be standing still; as they produce more and more data about us, we aren't getting any more intelligent; as they become more and more powerful, we might even be losing control of our own lives. Instead of the Singularity, we may actually be on the brink of its antithesis—let's call it the "Duality"—an ever-deepening chasm between humans and smart machines and also between tech companies and the rest of humanity.

Yes, Dov Seidman is right. Moore's Law is, indeed, unmooring us. It feels as if we are drifting toward a world that we neither quite understand nor really want. And as this sense of powerlessness increases, so does our lack of trust in our traditional institutions. The 2017 Edelman Trust Barometer, the gold standard for measuring trust around the world, recorded the largest-ever drop in public trust toward public institutions. Trust in media, government, and our leaders all fell precipitously across the world, with trust in media being, for example, at an all-time low in seventeen countries. According to Richard Edelman, the president and CEO of Edelman, the implosion of trust has been triggered by the 2008 Great Recession, as well as by globalization and technological change.[11] This trust scarcity is the "great question of our age," Edelman told me when I visited him at his New York City office.

It seems paradoxical. On the one hand, the digital revolution certainly has the potential to enrich everyone's life in the future; on the other, it is actually compounding today's economic inequality, unemployment crisis, and cultural anomie. The World Wide Web was supposed to transform mankind into One Nation, what the twentieth-century Canadian new media guru Marshall McLuhan called, not without irony, a global village. But today's Duality isn't just limited to the chasm between humans and computers—it's also an appropriate epithet for the growing gap between the rich and the poor, between the technologically overburdened and the technologically unemployed, between the analog edge and the digital center.

The Map Is the Message

Just as at other radically disruptive moments in history, we are living simultaneously in the most utopian and dystopian of times. Technophiles promise us an abundant digital future; Luddites, in contrast, warn of an imminent techno-apocalypse. But the real problem lies with ourselves rather than with our new operating system. So the first step in fixing the future is to avoid the trap of either idealizing or demonizing technology. The second step is much trickier. It's remembering who we are. If we want to control where we are going, we must remember where we've come from.

There's one more paradox. Yes, everything might seem to be changing, but in other ways, nothing has really changed at all. We are told that we are living through an unprecedented revolution—the biggest event in human history, according to some; an existential threat to the species, according to others. Which may be true in some senses, although we've heard the same sort of dire warnings in the past. Back in the nineteenth century, for example, similar warnings were made by romantics like the poet William Blake about the catastrophic impact on humanity of what he called the "dark Satanic mills." The future has, indeed, been both broken and fixed many times before in history. That's the story of mankind. We break things and then we fix them in the same way that we always have—through the work of legislators, innovators, citizens, consumers, and educators. That's the human narrative. And the issues that have always been most salient during previous social, political, and economic crises—the exaggerated power and wealth of elites, economic monopolies, excessively weak or strong government, the impact of unregulated markets, mass unemployment, the undermining of individual rights, cultural decay, the disappearance of public space, the existential dilemma of what it means to be human—are the same today as they've always been.

History is, indeed, full of such moments. In December 1516, for example, a little book was published in Louvain, today a university town in Belgium, then part of the Spanish Netherlands. This book came into a world that was in the midst of even more economic disruption and existential uncertainty than our own. The assumptions of the traditional feudal world were being challenged from every imaginable angle. Economic inequality, mass unemployment, and a millenarian angst were all endemic. The Polish astronomer Nicolaus Copernicus had just stumbled on the almost unspeakable realization that our planet wasn't the center of the universe. The democratizing technology of Johannes Gutenberg's printing press was undermining the centuries-old authority of the Catholic clergy. Most disorientating of all, populist preachers such as Martin Luther had invented the terrifying new theology of predestination that presented a Christian God of such infinite and absolute power that humans no longer had any free will or agency to determine their own fates. For many sixteenth-century folk, therefore, the future appeared profoundly broken. New cosmology and theology seemed to have transformed them into footnotes.

They couldn't imagine a place for themselves, as masters of their own destiny, in this new world.

That little book might, in part at least, have been intended to fix the future and reestablish man's confidence in his own agency. It wasn't much more than a pamphlet, written by a persecutor of heretics and a Christian saint, a worldly lawyer and an aspiring monk, a landowner and the conscience of the landless, a vulgar medieval humorist and a subtle classical scholar, a Renaissance humanist and a hair-shirted Roman Catholic, someone who was both an outspoken defender and an implicit critic of the old operating system of sixteenth-century Europe.

His name was Thomas More, and the book, written in Latin, was called *Utopia*—which can be translated into English as "No Place" or "Perfect Place." More imagined an island outside time and space, a simultaneously dreamlike and nightmarish one-nation kind of place featuring a highly regulated economy, full employment, the complete absence of personal privacy, relative equality between men and women, and an intimate trust between ruler and ruled. In More's *Utopia*, there were no lawyers, no expensive clothes, no frivolities of any kind. This no-place was—and still is—a provocation, a place forever on the horizon, an eternal challenge to the establishment, the most seductive of promises, and a dire warning.

Today, on its five-hundredth anniversary, we are told that this idea of Utopia is making a "comeback."[12] But the truth is that More's creation never truly went away. *Utopia*'s universal relevance is based on both its timelessness and its timeliness. And as we drift from an industrial toward a networked society, the big issues that More raises in his little book—the intimate relationship between privacy and individual freedom, how society should provide for its citizens, the central role of work in a good society, the importance of trust between ruler and ruled, and the duty of all individuals to contribute to and improve society—remain as pertinent today as they've ever been.

The Irish playwright Oscar Wilde captured this timelessness in 1891 when discussing the then-new operating system of industrial capitalism. "A map of the world that does not include Utopia is not even worth glancing at, for it leaves out the one country at which Humanity is always landing. And when Humanity lands there, it looks out, and, seeing a better country, sets sail," Wilde wrote in "The Soul of Man Under Socialism," his moral critique of what he considered to be the immoral factories and slaughterhouses of industrial society.[13]

So what was the core message—"More's Law," so to speak—buried in this enigmatic sixteenth-century text?

It's a question that has preoccupied generations of thinkers. Some see More as being nostalgic for a feudal commons that protected the so-called commonwealth of the traditional medieval community. Progressives like Oscar Wilde view the little book as a moral critique of nascent capitalism, while conservatives see it as a savage satire of agrarian communism. And then there are those—remembering More's close friendship with the Dutch humanist theologian Erasmus of Rotterdam, the seriocomic author of *In Praise of Folly*—who see the book as little more than an extended practical joke, the cleverest of humanist follies.

All these different interpretations have sought clues in a text that is defiantly elusive. But there is another, quite different, way of looking at it. There were four editions of *Utopia* published between 1516 and 1518, the first in Louvain, the second in Paris, and the third and fourth—which, according to historians, were closest to More's intent—in the Swiss city of Basel.[14] The most striking difference between the first and last versions lies in the imaginary map of Utopia that visualizes the invented island. The Basel editions contain an elaborate map, commissioned by Erasmus and designed, most likely, by the Renaissance artist Hans Holbein the Younger, who is best known now for his 1533 painting *The Ambassadors*, a humanist masterpiece that, in its surreal dissonance, captures the sense of crisis pervading the period.[15] Holbein also painted Thomas More's portrait in 1527—a more personalized masterpiece that captured the surreal dissonance in More's life between a man of the world and a man of God.

This map may be the message. At first glance it appears to be of a hilly, circular island with a fortified town at its center and a harbor in the foreground sheltering two anchored ships. A closer examination, however, reveals a very different kind of geography. By closing one eye and staring at the illustration slightly off-kilter, we see Utopia transformed into a grinning human skull, a symbol denoting *memento mori*, the Latin expression meaning "Remember you have to die," and a familiar trope in both classical Rome and medieval Europe. The island itself represents the skull's outline. One ship is the neck and an ear; the other ship is the chin, with its mast as the nose and its hull as the teeth. The town is the forehead, with a combination of the hills and the river being the eyes of the skull.[16]

So what, exactly, was the point of transforming the map of Utopia into the image of a skull? As so often with More and his early-sixteenth-century humanist friends, there's an element of esoteric humor here, with *memento mori* being a play on More's surname and the substitution of the island for a skull representing a classic Erasmian folly. But there's another, more life-affirming message, which, like the outline of that skull hidden in the map, isn't immediately obvious to the eye.

The great debate of the early sixteenth century was between Renaissance humanists, such as More and Erasmus, and Reformation preachers like Luther, and it addressed the question of free will. Luther, you'll remember, in his theory of predestination, presented a God of such absolute power that humans were shorn of their agency. The humanists, however, clung to the idea of free will. More's *Utopia* is, indeed, a manifestation of that free will. By inventing an ideal society, More was demonstrating our ability to imagine a better world. And by presenting his vision of this community to his readers, he was inviting them to address the real problems in their own societies.

Utopia, then, is a call to action. It assumes that we possess the agency to improve our world. Therein lies the other significance of that grinning skull in Holbein's map. In ancient Rome, the expression *memento mori* was used to remind successful generals of their fallibility. "*Memento mori ... Respice post te. Hominem te esse memento,*" the slaves would shout at the triumphant general during the public parade after a great military victory. "Yes, you will die," the slave reminded the Roman hero. "But until then, remember you're a man." In pagan Rome, then, that skull was as much a symbol of life as of death. It was a reminder to cultivate the civic self and to make oneself useful in public affairs while one still had the chance.

In contrast with the technological determinism of Moore's Law, this law of More's refers to our duty to make the world a better place. In Utopia too, there is much talk of the "duty" we ought to have to our community. "All laws are promulgated for this end," More writes, "that every man may know his duty."

More's Law, then, is Thomas More's definition of what it *should* mean to be a responsible human being. He not only tried to live his life according to this principle; he also died from it, beheaded by his king, Henry VIII, for refusing to sanction Henry's divorce from his first wife. Being part of the human narrative, More believed, means seizing control of our civic and secular fate.

In today's age of acceleration, five hundred years after the publication of *Utopia*, many of us once again feel power-less as seemingly inevitable technological change reshapes our society. As More reminds us, fixing our affairs—by becoming steersmen or pilots of society—is our civic duty. It's what made us human in the sixteenth century, and it's what makes us human today.

"*Hominem te esse memento,*" the Roman slave would remind the victorious general. As we drift with seeming inevitability into a new hyperconnected world, these are words we should remember as we fight to establish our place in this unfamiliar landscape.

Humanity Is Trending

In Thomas Friedman's 2016 bestselling *Thank You for Being Late: An Optimist's Guide to Thriving in the Age of Accelerations*, there is a fifty-page introductory chapter lauding "Moore's Law" and all but designating it the foundational truth about early-twenty-first-century society.[17] But Gordon Moore's observation about the processing power of silicon chips isn't a particularly helpful guide, for either optimists or pessimists, to thriving in an age of accelerations. As Dov Seidman reminds us, it doesn't tell us who we really are as human beings.

More's Law is more useful because it explains how we should fill that hole in the future with human agency. "Humanity" is now trending in tech. It might not be quite as Manichaean a showdown as the "Technology Versus Humanity"[18] or the "Digital Versus Human"[19] cage match foreseen by some futurists, but the human costs of the digital revolution are quickly becoming the central issue of our digital society. Everyone, it appears, is waking up to a confrontation that the Israeli historian Yuval Noah Harari frames as "Dataism" versus "Humanism"—a zero-sum contest, he claims, between those who are known by the algorithm and those who "know thyself."[20] And everyone seems to have his or her own fix to ensure that "Team Human,"[21] as the new media guru Douglas Rushkoff puts it, wins.

Everyone, it seems, wants to know what it means to be human in the digital age. A few days before the "Encrypted and Decentralized" event at the Alte Teppichfabrik, for example, I participated in a lunch discussion in Berlin that was unappetizingly called "Toward a Human-Centered Data Revolution." The month before, I'd spoken at Oxford about "The True Human," in Vienna about "Reclaiming Our Humanity," and in London about why "The Future of Work Is Human." Klaus Schwab, the Swiss founder of the World Economic Forum, exemplifies this preoccupation with a new humanism. It "all comes down to people and values," he explains about the impact of digital technology on jobs,[22] which is why we need what he calls "a human narrative" to fix its problems.[23]

To write a *human* narrative in today's age of smart machines requires a definition of what it means to be human. "As soon as you start defining the question *what is human*, it becomes a belief," the writer and inventor Jaron Lanier once warned me over lunch in New York as we prepared for a debate about the impact of AI on humanity. Lanier may be right. But in a world in which we've invented technology that is *almost* human, it seems only natural that we would want to compare ourselves with smart machines in an effort to define both ourselves and this new technology. Besides, if we don't believe in our own humanity, then what *can* we believe in?

To come up with a distinction between human beings and computers, I spoke to Stephen Wolfram, the CEO of the Massachusetts-based computer software company Wolfram Research and one of the world's most accomplished computer scientists and technology entrepreneurs. Educated at Eton, Oxford, and Caltech, Wolfram was awarded his doctorate in theoretical physics at the age of twenty and a MacArthur Fellowship at twenty-two, the youngest person ever to have received one of these $625,000 "genius" awards. He is the author of the bestselling and critically acclaimed *A New Kind of Science*. He's the creator of the influential mathematical software program Mathematica and the curated online knowledge resource WolframAlpha, a kind of superintelligent Google, which, among many other things, is the engine providing the factual answers to queries submitted to Apple's Siri. And, as if all that weren't enough, he's the inventor of the Wolfram Language, a programming language built on top of Mathematica and WolframAlpha that is designed to help us communicate with computers.

I first met Wolfram in Amsterdam at the Next Web Conference. But rather than discussing the abstract future, we spent a most pleasant evening chatting about our personal futures—our own children. He is a champion of homeschooling, with a couple of his kids being educated at home by him and his mathematician wife. His mother, Sybil, was a teacher too—an Oxford University philosopher with a particular interest in Ludwig Wittgenstein's philosophy of language.

"What do I *do*?" Wolfram repeats my words gingerly, as if nobody had ever asked the multimillionaire software entrepreneur, world-famous physicist, and bestselling writer such a challenging question.

He explains that what he does—or at least tries to do— is teach humans to understand the language of machines. He is building an AI language we can all understand.

"I want to create a common language for machines and people," he tells me. "Traditional computer languages pander to machines. While natural human language isn't replicable in machines."

I ask him if he shares the fear of AI pessimists who believe that technology could develop a mind of its own and thereby enslave us.

Computers, the thinking machines imagined by the Victorian mathematician Ada Lovelace and her business partner, Charles Babbage, in the mid-nineteenth century, are the defining invention of the last couple of hundred years, Wolfram explains. But the one thing that they don't possess, he insists, is what he calls "goals." Computers don't know what to do next, he says. We can't program them to know that. They couldn't write the next paragraph of this book. They can't fix the future.

Wolfram is a great admirer of Ada Lovelace, and his argument is, essentially, a rephrasing of her thoughts on the intellectual limitations of computer software. "The Analytical Engine has no pretensions whatever to originate anything. It can do whatever we know how to order it to perform . . ." Lovelace famously wrote in 1843. "Its province is to assist us in making available what we are already acquainted with."[24]

"If a lion could speak, we could not understand him," Wolfram says, quoting one of Ludwig Wittgenstein's most elliptical aphorisms from his *Philosophical Investigations*. And the same is true of even the smartest thinking machines, he says. If one of these machines could talk, we couldn't understand its real meaning, because of our differences. If machines could speak our language, they wouldn't be able to fully understand us, because we have goals and they, as Ada Lovelace explains, can't "originate" anything.

The truth about the meaning of "humanity" is that there's no truth. No absolute truth, at least. Every generation defines it according to its own preoccupations and circumstances. So, for example, the humanism of the original Renaissance was rooted in discovering and then reconnecting with a history that had been lost in the Dark Ages. For Thomas More or Niccolò Machiavelli, being human meant putting on the robes, sometimes quite literally, of antiquity. Five hundred years later, our preoccupations and circumstances are very different. What it means to be human today is bound up in our relationship with networked technology, particularly thinking machines. If there is to be a new renaissance, this relationship with smart tech will be the core of its new humanism.

Wolfram's definition, with its focus on human volition, is both timely and timeless. Our unique role in the early twenty-first century is, in Ada Lovelace's words, to be able to "originate" things. This is what distinguishes us from smart machines. But it's also an updated version of More's Law, with its reminder of our mortality and its focus on our civic responsibility to make the world a better place.

The way to solve the most vexing problems of the future, the WEF CEO Klaus Schwab says, is by creating a story about people—a *human* narrative. And that's the goal of this book too. In the midst of today's great digital transformation, this story features the many solutions of many different people in many different places to the many challenges of our new network epoch. They are all filling that hole in the future. Obeying More's Law, they are trying to design a new operating system for humans rather than for machines. What unites them all is their insistence that, in the age of the smart networked machine, we humans must seize back control of our own fate and, once again, author our own story.

Chapter Two

FIVE TOOLS FOR FIXING THE FUTURE

In Betaland

The nineteenth-century neighborhood is full of twenty-first-century things. I'm with my old friend John Borthwick, the founder and CEO of Betaworks, a New York City–based venture studio that incubates technology start-up companies. We are at the Betaworks studio in New York's Meatpacking District—the downtown Manhattan neighborhood named after its industrial-scale slaughterhouses—which is now one of the city's most fashionable areas. Along with its cobbled streets, boutique stores, exclusive clubs, and restaurants, the area is best known as being the southern terminus of the High Line—the section of the old New York Central Railroad that has been successfully reinvented as a three-mile-long elevated public park.

Borthwick's studio is located in a cavernous old brick building that has been converted from a decaying warehouse into an open-plan workspace. The place is lined with young computer programmers—Betaworks' so-called "hackers in residence"—peering at electronic screens. It's a kind of renaissance. The analog factory has been reborn as a digital hub. These hackers are manufacturing the twenty-first-century networked world from inside a nineteenth-century industrial shell.

But this new world is still in *beta*—the word the tech industry uses to describe a product that's not quite ready for general release. And it's this emerging place—*betaland*, so to speak—that I've come to talk about with Borthwick. We've been friends for years. Like me, he was a start-up entrepreneur during the first internet boom of the mid-nineties. In 1994, fresh out of Wharton business school, he founded a New York City information website called Ada Web, in honor of Ada Lovelace. Borthwick sold Ada Web and several other internet properties to the internet portal America Online in 1997 and became their head of new product development. He then ran technology at the multinational media conglomerate Time Warner before founding Betaworks in 2008, and there he's made his fortune investing in multibillion-dollar hits such as Twitter and Airbnb.

"I fell in love with the idea of the internet," Borthwick says, explaining why he became an internet entrepreneur, articulating the same faith as such mid-twentieth-century pioneers as Norbert Wiener that networked technology could pilot us to a better world. It was the *idea* that a new networked world could be better than the old industrial one. The *idea* that the internet could transform society by making it more open, more innovative, and more democratic.

Over the last quarter century, however, Borthwick's youthful faith in this idea has evolved into a more ambivalent attitude toward the transformative power of digital technology. As we sit in one of the studio's meeting rooms, surrounded by his hackers in residence, we speculate on the networked world on the horizon. The innocence of the nineties, the faith in the internet's seemingly unlimited potential—all that openness, innovation, and democracy—has been replaced by the realization that things aren't quite right in *betaland*.

As we talk, we realize we agree that today's vertiginous atmosphere of social divisiveness, political mistrust, economic uncertainty, and cultural unease is—in part, at least—a consequence of the digital revolution. In contrast, however, with the crusading Edward Snowden, Borthwick is realistic rather than pessimistic about the future. He understands as well as anyone the remarkable achievements of the digital revolution, but he is cognizant of the problems too. He is—like me—a *maybe*.

So, how to rebuild the future and manifest the human agency that Snowden says we've lost? "Five fixes, John," I say. "Give me five bullet points on how we can fall back in love with the future."

Borthwick's Five Bullets

A cheerful, boyish-looking fellow with a mop of dark hair, Borthwick grins at my challenge. Rather than the personal computer or the internet, for him the new *new* thing is artificial intelligence—the technology behind networked smart machines, smart cars, smart algorithms, smart homes, and smart cities. It's the superintelligent technology that some people fear could destroy humanity, and that's what preoccupies Borthwick at first when he addresses my question.

"Look, so much of our business is reality distortion," he confesses about the tech industry, "so we don't fucking know what AI is going to be."

But he does know what AI *shouldn't* be—a proprietary operating system owned and operated by a single winner-take-all company. So his first fix is what he calls an "open AI platform," a public space for technologists, not unlike the Meatpacking District's High Line—the public park built on top of the New York Central railroad that was created by an alliance of local entrepreneurs, city regulators, and urban activists. Borthwick says it's critical that what he calls "the stack"—the manifold layers of technology making up a networked operating system—is open to every type of developer and application. To borrow some language from the Berlin venture firm BlueYard Capital, Borthwick wants to "encode" the "value" of openness into the architecture of the internet. It's a kind of network neutrality for the AI age. And his model for this is the World Wide Web, the open platform so generously donated to the tech community by Tim Berners-Lee in 1989, on which innovative first-generation internet companies like Skype, Amazon, and Borthwick's own Ada Web flourished. And that's why, Borthwick tells me, he is advising the nonprofit Knight Foundation on its "Ethics and Governance of Artificial Intelligence Fund"—a

$27 million fund announced in 2017 that is dedicated to researching artificial intelligence for the public interest.[1]

But as Borthwick acknowledges, for every public-spirited Berners-Lee or Knight Foundation, there is a private corporation seeking to dominate the market through its complete control of the technology stack. Hence the need for antitrust regulation—his second fix. As the head of new products at America Online, Borthwick was involved in the US government's 2002 labyrinthine antitrust case against Microsoft, and he is under no illusions about the cost, in both time and money, of battalions of high-priced lawyers squabbling over the legal and technological minutiae of computer operating systems.[2] And yet, as an investor in early-stage innovation, he understands the need to protect his hackers in residence from contemporary leviathans like Google, Amazon, and Apple. Antitrust matters to investors in innovation such as Borthwick. Start-up entrepreneurs and technologists, he says, need the protection of government against winner-take-all multinationals. Regulation, the venture capitalist suggests, sometimes *is* essential to protect innovation.

His third fix also focuses on the importance of the public sphere. As an early investor in Twitter, he is acutely aware of the social media company's struggle to find a compelling business model as well as its increasingly central role in our culture and in making news. Twitter's value, he suggests, is more than just economic—particularly in our "post-truth" Trumpian political world, with its infestation of fake news and its nefarious trolls whipping up the online mob.[3] So, he says, we should approach prominent media companies like Twitter as if they are public television or radio networks.

"It's important to run Twitter on the same public principles as Harvard University or the *Guardian* newspaper. Its societal worth can't be quantified purely in terms of financial value," the Betaworks CEO insists. What he's saying is that—given the corrosive impact of fake news and other ridiculously biased or corrupt information on a culture that is already suffering a dearth of trust—some of our most influential new media platforms, like Twitter, might be too *important* to be treated like other for-profit companies.

Borthwick's next fix focuses on the increasingly blurred distinction between people and smart machines, particularly in an age of augmented reality, when it will be increasingly hard to distinguish between a human body and networked devices. The venture capitalist shuffles in his seat and addresses the existential dilemma at the heart of smart technology.

"At what point are we no longer human?" he asks tentatively. It sounds as much a plea as a question. Today's technology, he says, comes with great moral responsibility. Such *great* responsibility, in fact, that we need to establish "criteria for human-centric design in order to retain our identity as a species." Borthwick doesn't explain where these "criteria" for determining humanness will originate. Maybe from the designers of today's existentially disruptive technology. Maybe from new governmental laws. Maybe from concerned citizens like Edward Snowden or public interest groups like the Knight Foundation. Maybe from investors like Borthwick himself.

But there's no absence of human agency, nothing quizzical about Borthwick's final bullet. When a Wharton graduate and successful venture capitalist approvingly quotes *The Communist Manifesto*, you know something very odd is going on in the world. Yet that's exactly where Borthwick begins his fifth fix. Roll back the clock to the middle of the nineteenth century, he says, and we would find a similar world of radically disruptive new technology, deep inequalities between rich and poor, grossly unsanitary working conditions, high unemployment, and monopolistic capitalist companies.

Borthwick's comparison with the industrial revolution is instructive. As Eric Hobsbawm, the Anglo-German historian of the industrial revolution, reminds us, the world in 1789 was "overwhelmingly rural" and for most of its inhabitants, who had no access to either newspapers or any other information about the outside world, "incalculably vast."[4] It was a localized, agricultural society that wouldn't have appeared strikingly foreign to Thomas More, who died more than two hundred years earlier. But sixty years later, by February 1848, when Karl Marx and Friedrich Engels published their *Communist Manifesto* in London, that rural world had been radically transformed by the steam engine, electrification, and mass mechanical production. "The mid-nineteenth century was pre-eminently the age of smoke and steam,"[5] Hobsbawm tells us about a world in which, between 1850 and 1870, global production of coal multiplied by more than 250 percent, global iron output grew by 400 percent, and total global steam power went up more than 450 percent.[6]

"All that is solid melts into air." In language akin to that used to describe the dramatic impact of Moore's Law on contemporary society, Marx and Engels described an industrial revolution where "constant revolutionizing of production, uninterrupted disturbance of all social conditions, everlasting uncertainty and agitation distinguish the bour-geois epoch from all earlier ones."[7]

The gaping inequalities and injustices of the mid-nineteenth-century industrial world have been well doc-umented by everyone from Marx and Engels to Thomas Hardy and Charles Dickens. The "exploitation," *The Com-munist Manifesto* claimed, was "naked, direct, shameless, bru-tal."[8] By 1848, the industrial revolution had wrought what the Hungarian economist Karl Polanyi—author of *The Great Transformation*, the classic 1944 account of the shift from an agricultural to an industrial economy—called "unprec-edented havoc with the habitation of common people."

In the mid-nineteenth-century English economy, Polanyi says, "the laboring people had been crowded together in new places of desolation, the so-called industrial towns of England; the country folk had been dehumanized into slum dwellers; the family was on the road to perdition; and large parts of the country were rapidly disappearing under the slag and scrap heaps vomited forth from the 'satanic mills.'"[9] The end result, he says, was the creation of "Two Nations": one of "unheard-of wealth," the other of "unheard-of poverty."[10]

Polanyi traces what he calls the "catastrophic disloca-tion of the lives of the common people" back to the same sixteenth-century enclosure movement in which capital-ist practices created mass agricultural unemployment that Thomas More critiqued in *Utopia*. "Your sheep that com-monly are so meek and eat so little," More wrote acidly about sixteenth-century England, "have become so greedy and fierce that they devour human beings themselves."[11] The problem with both enclosures and early industrialization,

Polanyi argues, is that a "common-sense attitude toward change was discarded in favor of a mystical readiness to accept the social consequences of economic improvement, whatever they might be."[12] The ideal of the free self-adjusting economic market, Polanyi says, "implied a stark utopia" that couldn't exist "for any length of time without annihilating the human and natural substance of society."[13] Rather than being against capitalism itself, Polanyi's attack was against a cult of the free market, which viewed any kind of regulation or interference in it as a fundamental assault on freedom.

Eric Hobsbawm echoes Polanyi's apocalyptic rendering of life in mid-nineteenth-century Europe. He explains that the most "elementary services of city life"—such as street cleaning or the water supply or basic sanitation—couldn't keep up with the dramatic economic and technological transformation. Thus there were periodic epidemics of cholera and typhus throughout the Europe of the 1830s to the 1850s. And mass alcoholism, Hobsbawm explains, created a "pestilence of hard liquor" across Europe.[14] What he calls a "social and economic cataclysm" spawned not only a plague of "infanticide, prostitution, suicide and mental derangement" in the cities,[15] but also the mass famines of 1795, 1817, 1832, and 1847 in the European countryside. Even places with full employment, Hobsbawm reports, were impoverished, with 52 percent of workers employed in a good trade in the Lancashire mill town of Preston living below the poverty line in 1852. And if you were one of the lucky few who survived till old age, Hobsbawm concludes, your life would be "a catastrophe to be stoically expected" since there was no government health care or social security system to support old people.[16]

I glance out the window of the Betaworks studio onto the cobblestoned Meatpacking District street and try to imagine the equivalent scene in February 1848 when Marx and Engels published their *Manifesto*. In the mid-nineteenth century, many New York slaughterhouses operated outside the law, running filthy factories that produced unsafe meat and "employing" impoverished workers without providing any job security. It was bedlam. Half of every cow slaughtered was inedible, and much of this foul-smelling animal waste ended up in local rivers and lakes.[17] A common practice was to stack the unused parts of animal carcasses outside the slaughterhouses till they were transported elsewhere. Children played in streets that ran with the blood of slaughtered animals. No wonder diseases like tuberculosis and diphtheria were so endemic in the city.

And it wasn't much better anywhere else in nineteenth-century industrial New York. In the second half of the century, the city's Lower East Side was, according to Hobsbawm, "probably the most overcrowded slum area in the western world with over 520 persons per acre."[18] Working conditions were no better, particularly in New York's garment factories. Laboring sixty-hour weeks, garment workers risked being killed by machinery that regularly severed limbs and ripped off scalps. According to the 1900 US Census, 6 percent of the workforce consisted of kids between the ages of ten and fifteen doing the work of adults. Riots against these working conditions were commonplace. And so were the attacks on and intimidation of rebellious workers by thugs hired by the factory owners.

Borthwick reminds me of what's broadly changed over this time in industrial society. "Fast-forward a century, and we've done the unthinkable. Eight out of ten of the goals of the *Manifesto* were peacefully realized," he says, referring to what Marx and Engels described as the ten "measures" that would create the conditions for a classless society in advanced industrial countries. These goals included the imposition of a progressive or graduated income tax, the creation of a national bank, the establishment of publicly owned land, government control over transportation and communications, free education for all children in public schools, and the abolition of child labor in factories.[19]

His point is that what seemed beyond reach to most people in 1850—banning child labor in factories and the creation of a national bank, for example, or free public education and a graduated tax system—is now taken for granted in almost every country in the world. These reforms weren't always ideally implemented, Borthwick admits, but over the last hundred and fifty years they've slowly established themselves as the essential foundations of a civilized society. So, for example, in the 1880s, thirty years after Marx and Engels published their *Manifesto*, Otto von Bismarck, the chancellor of the newly established united Germany, introduced a mandatory national program of social welfare that included health, accident, and old-age insurance for industrial workers. These reforms were then developed and adapted by every industrializing country to conform to its own political culture—from the German corporatist model to the Scandinavian social democratic tradition to the more market-based social security systems of Anglo-Saxon countries like the United Kingdom and the United States.

Today, Borthwick explains, we are faced with a new set of dramatic challenges that may require similarly *unthinkable* solutions. No, he acknowledges, the digital revolution hasn't re-created the "satanic mills" of Northern England, with its armies of child laborers and appalling working conditions. Nor do the streets of New York City run anymore with the blood of dead animals. History never repeats itself. Not exactly, anyway. But the great transformation of the early twenty-first century is, nonetheless, equally challenging— particularly in terms of economic inequality and the future of employment. Karl Polanyi's Two Nations are back. Today, however, these nations are being shaped by digital technology rather than by steam or electricity. Today, it's smart machines—rather than Thomas More's enclosed farmland or Marx and Engels's mechanized factories—that are radically disrupting not only employment but the very nature of working life itself. A 2013 Oxford University white paper, for example, forecasts that 47 percent of jobs could be eliminated by smart technology over the next two decades,[20] and a 2017 McKinsey & Company report predicts that 49 percent of all the time we spend working could be automated by current technology.[21]

And so, on to Borthwick's fifth and most ambitious fix. We need, he tells me, to create completely new entitlement programs, educational institutions, and social security systems designed for today's networked society. Just as the nineteenth-century industrial revolution resulted in the reinvention of the relationships between employees, employers, and the government, he says, so today's digital revolution needs to spark the same radical rethinking. One early example of this, Borthwick says, is what's known as the "minimum guaranteed income"—an idea designed to provide a minimal amount of economic security at a time of technology-driven unemployment. "We can't have a stable society," he insists, "unless you have a fair part of the community employed."

So that is John Borthwick's manifesto, his own five bullets to fix the digital future:

- Open technology platforms
- Antitrust regulation
- Responsible human-centric design
- The preservation of public space
- A new social security system

But none of these bullets is magical, he explains. In comparison, for example, with many late-nineteenth- and early-twentieth-century Marxists, who believed that every problem of industrial capitalism would be instantly fixed by a global proletarian revolution, Borthwick is under no illusion that any single reform will automatically solve all the problems of the great digital transformation. There is no grand Hegelian synthesis in his narrative. No end of history. No island of Utopia on the horizon.

Borthwick's vision of the future is, to borrow a word from Immanuel Kant, *crooked*. He uses the metaphor of the stack—the layers of technology supporting an effective platform—to explain how all his disparate ideas could fit together in the future. He is suggesting that they are all interchangeable pieces of a new twenty-first-century operating system, one that is constantly evolving and adapting to fresh technological innovation.

Yes, he acknowledges, the combinatorial effects of the internet, cloud and mobile technology, artificial intelligence, and big data are dramatically changing the world. In response, the venture capitalist says, we need to foster equally powerful combinations of public policy, ethical responsibility, legal reform, and technological innovation. These reforms shouldn't exist in silos, isolated from one another, he insists. Like the stack of products supporting a successful technological platform, they are most effective when working in combination with each other.

A Combinatorial Strategy

What is striking about Borthwick's five bullets is that, in spite of the radically disruptive nature of digital technology, none of his points are new. From English cotton mills to Manhattan slaughterhouses, from the inequities of nineteenth-century Lancashire mill towns to those of nineteenth-century New York City, from the enclosures of Thomas More's sixteenth-century England to the Two Nations of Marx and Engels's nineteenth-century Europe, we've seen this show before.

Not identically, of course. As I just noted, history never repeats itself exactly. But many of the problems of the industrial past resemble many of those now confronting us in our digital present. And so the essence of John Borthwick's tools for solving these problems—the civic value of openness, the need for public space, the symbiosis of innovation and regulation, and, of course, the need to reinterpret what it means to be human in times of radical technological upheaval—are anything but new.

Nor are the strategies new for achieving this change. There are, after all, only so many peaceful ways of going about solving the world's problems. This book focuses on what, I think, are the five perennial methods for fixing the future.

The first category, as John Borthwick suggests, is through legal regulation—the establishment, for example, of antitrust regulations to protect competition or the creation of public media to guarantee the free flow of reliable information. Hard-core free market libertarians, both inside and outside Silicon Valley, will strongly disagree, of course. But as we will show throughout this book, particularly in Chapter Six, they are wrong. Innovation and regulation are symbiotic, and the future never gets fixed without new laws. As Thomas More reminds us, good government has been critical throughout human history. That will never change.

The second method is through the work of such innovators as Betaworks' hackers in residence or the entrepreneurs at BlueYard Capital's "Encrypted and Decentralized" event, who invent new technologies and products to improve people's lives. This doesn't mean that every innovator or innovation is good. As we shall see in the next chapter, much of the digital innovation of Big Tech companies like Google and Facebook isn't currently working. Yet, as I show in Chapter Seven, that isn't an excuse to write off innovative entrepreneurs altogether in terms of creating products that realize the promise of the digital revolution. Utopians might dream up economic systems theoretically superior to free market capitalism, but none exist in the real world.

The third is through the behavior of consumers, whose choices on what they want and how much they are willing to pay for it reshape markets and products. The consumer isn't always right, of course, particularly in our digital age of absurdly low-priced or "free" products. But as we shall see later in this section, with the history of the food industry, it has been consumers who have played a central role in fixing many of the early problems of industrial capitalism by demanding higher quality and healthier products.

The fourth is through the initiatives of trade unions, philanthropists, nonprofits, or committed individuals like Edward Snowden. Strategies to realize this, as I show in Chapters Eight and Nine, include strikes and other direct labor action, public work, donations, and pressure on both government and industry. This is the category most singularly driven by More's Law—by people's will to improve the lives of their fellow citizens.

The fifth is the role of education—from parents, teachers, mentors, policy makers, even books like this— in helping people shape their own histories and determine the best outcomes for their society. As we will see in Chapter Ten, the reinvention of education is particularly challenging. It requires not only new kinds of teachers and schools, but also a new way of thinking about the purpose of education itself.

These, then, are our five tools for fixing the future:

- Regulation
- Competitive Innovation
- Social Responsibility
- Worker and Consumer Choice
- Education

Betaworks CEO John Borthwick is right. There isn't a single magic bullet to fix the future. But, as this book will demonstrate, these five reform strategies—when implemented creatively—can together transform betaland into a better land. What's essential is that they exist together—as part of a stack enabling a new operating system for twenty-first-century society. Things go wrong when one of these tools is particularly fetishized and thus dominates all the others. So, for example, the cult of the perfectly unfettered market, Karl Polanyi's "stark utopia," caused many of the problems of nineteenth-century industrial capitalism. But, of course, the Soviet response to Two-Nation industrial capitalism, with its cult of governmental power and banning of any kind of competitive innovation, was even more flawed.

The key, then, is a combinatorial strategy. Throughout history it's been these five broad tools—regulation, innovation, worker and consumer choice, social responsibility, and education—that have, one way or the other, combined to manage dramatic societal change. This has been particularly true over the last couple of hundred years, as the excesses and injustices of a nascent industrial society have mostly been vastly improved by a tangled combination of these five kinds of action.

As it happens, the American meatpacking industry is a good example of how these tools can combine to improve the world. Public outcry about unsafe and unclean production practices and adulterated meats, notably the scandal of the "embalmed" meat served to soldiers in the Spanish-American War, led to the publication of Upton Sinclair's muckraking novel *The Jungle* and to Teddy Roosevelt's signing of the Meat Inspection Act of 1906. This act, among other sanitary provisions, made government inspection mandatory. Since the latter half of the nineteenth century, civic groups like New York's Ladies' Health Protective Association, local agencies like the New York City Metropolitan Board of Health, and federal regulators had fought a recalcitrant slaughtering and packing industry that refused to see the benefits to itself in changing its business model. Attempts to secure voluntary cooperation failed time after time, with small butchers and big packers alike claiming that their trade posed no threat to the public. Slaughterhouses even protested prohibitions against driving cattle through the streets of the city. When regulation came, rather than hampering the business, it actually sped up the rate at which the industry modernized. The hard-won success of the body of reforms developed over the late nineteenth and early twentieth centuries ended up pushing meatpackers to innovate in technology, in design, and in production efficiency, all of which made meat safer and more affordable for average consumers. It ended up being a win for both producer and consumer.

The same is true more broadly, in terms of the role of our five principles, in the dramatic improvement in the quality of New York's living and working conditions. State workers' compensation laws and the 1935 National Labor Relations Act now protect workers' right to unionize, and strong unions have resulted in higher pay and safer workplaces. The 1970 Occupational Safety and Health Act reduced on-the-job deaths to 3.3 per 100,000 in 2013. The Fair Labor Standards Act of 1938 banned child labor, and today's better-educated workforce is due in part to that legislation, as well as to laws making school attendance mandatory. Most American adults over twenty-five today have completed high school; in 1930, less than 20 percent had. After hundreds died in New York's smog crises in the 1950s and '60s, clean air activists got polluting incinerators shut down. Federal regulations and the creation of the Environmental Protection Agency in 1970 reduced New York's acid rain and the amount of sewage in the Hudson River to almost nil. The average New Yorker's environmental footprint today is, in fact, smaller than that of a rural or suburban resident.

Most of all, the effectiveness of our combinatorial strategy is demonstrated in the story of the food industry during the industrial revolution. It's a narrative that parallels the development of progressive-era ideas in public health, social welfare, and the expansion of public education. Over the last hundred and fifty years, a combination of government regulation, competitive innovation, social engagement by business leaders, and consumer choice and education has resulted in the dramatic improvement in the quality, healthfulness, and even price of much of the food products available. What's emerged is a stack of new policies, organizations, and habits that have transformed a dysfunctional industry into one that is serving the interests of entrepreneurs, consumers, and society.

The mid-nineteenth-century economy of satanic mills produced equivalently satanic food. In 1845, Friedrich Engels argued in his encyclopedic *The Condition of the Working Class in England* that the mass-produced food sold to the workers was rife with adulterants. "The refuse of soap-boiling establishments also is mixed with other things and sold as sugar," Engels reported. "Chicory and other cheap stuff is mixed with ground coffee, and artificial coffee beans with the unground article. Cocoa is often adulterated with fine brown earth, treated with fat to render it more easily mistakable for real cocoa."[22]

The same was true with many mass-produced foodstuffs. "Smouch"—dried leaves of the ash tree used as filler in tea— was used so extensively that the British Parliament passed a law forbidding it in order to protect the ash tree! Manufacturers colored tea and cheese with deadly toxins like lead and verdigris. Sawdust was frequently included in foodstuffs to increase profits. In America, the situation was similar, with both an absence of regulation and the adulteration of many food products. By the beginning of the twentieth century, with the rise of such mass retailers as the A&P chain, many of the major new food brands—to preserve both taste and shelf life—became overprocessed and dominated by additives. Food may no longer have been satanic, yet, reflecting the increasingly homogenized and corporatized society of the time, it was bland, uniform, and mostly unhealthy.

The reaction to all this took time to develop. In the nineteenth century, there were very few US federal laws that regulated the contents or sale of foodstuffs. This changed with the 1906 Pure Food and Drug Act and the 1906 Meat Inspection Act, which led to the creation of the US Food and Drug Administration (FDA), an agency designed to protect consumers from harmful products. Over the last hundred years, the FDA has regulated dyes and chemicals in foodstuffs, ensured the honest labeling and marketing of products, and even pioneered what became known as the "poison squad" of volunteers to test food preservatives. In 1938 Franklin Delano Roosevelt signed the new Food, Drug and Cosmetic Act—greatly expanding the enforcement powers of the FDA and establishing new regulatory standards for foods—which still remains the legal foundation of the FDA's authority. Today, the FDA has a budget of $4 billion and employs fifteen thousand people.

A diversity of institutions and individuals—including nonprofits like the Rockefeller Foundation and the World Health Organization, as well as scientists at the University of California—pioneered research into the harmful effects of added sugar in foods. Education, particularly through books, was also important in terms of alerting people to the dangers of industrially processed foods. Arthur Kallet and F. J. Schlink's 1933 bestselling *100,000,000 Guinea Pigs: Dangers in Everyday Foods, Drugs, and Cosmetics*, which went through thirteen printings in its first six months of publication, first alerted Americans to the extent of contamination in their mass-produced foods. The release of Rachel Carson's 1962 bestseller *Silent Spring* represented a watershed moment in raising both public and government consciousness about the dangers of pesticides and toxic chemicals in the food and water supply. Later books such as Eric Schlosser's 2001 *Fast Food Nation: The Dark Side of the All-American Meal*, which was also made into a 2006 movie, alerted people to many of the most troubling health and environmental effects of the fast-food industry.

While obesity and poor eating habits remain a major problem in America, there has been a significant shift over the last quarter century to a much healthier food culture. This has been driven by a symbiosis of consumer demand and entrepreneurial innovation. When John Mackey opened the first Whole Foods Market in Austin, Texas, in 1980, there were only a handful of quirky natural foods supermarkets in America selling produce without artificial preservatives, colors, flavors, sweeteners, and hydrogenated fats. By June 2017, Whole Foods, now a Fortune 500 public company with 431 stores and ninety-one thousand employees, was acquired by Amazon for $13 billion. Other successful chains of natural supermarkets include Trader Joe's, Sprouts, and Earth Fare. Even mainstream grocery chains like Safeway have introduced organic produce sections in all their stores, and Safeway has a "100 percent natural" Open Nature brand. The last couple of decades have also seen the growing popularity of farmers' markets directly selling local products. In 1994 there were 1,755 farmers' markets in the United States. Today there are 8,600 of these markets generating annual revenues of more than $1.5 billion.

What has happened over the last century in this decisive shift to natural foods, higher-quality supermarkets, and healthier eating habits represents a model for fixing our networked future. Like the food industry in 1850, today's digital economy is characterized by an uncontrolled free market, addictive products, corporate irresponsibility, and wide-scale ignorance of the impact of this technology on our mental and physical well-being. The success of the food industry over the last few decades in reinventing itself proves that the combinatorial effects of responsible regulation, innovation, worker and consumer choice, citizen activism, and education can make possible what John Borthwick calls the "unthinkable."

But before we get to the stack of solutions that can fix the digital revolution, we still need more detail on what's gone wrong with this great transformation. Karl Polanyi, you'll remember, described the mid-nineteenth-century industrial economy as "a stark utopia" annihilating society. The next chapter examines the equivalent utopian elements of today's networked economy. It explains how, exactly, the future is broken.

Chapter Three

WHAT IS BROKEN

Darkness on the Edge of Town

The future sometimes appears in the unlikeliest of places. It is midwinter, and I am in Tallinn, the capital of Estonia, the tiny Baltic republic on the northeastern edge of Europe. With the most start-ups per person, the zippiest broadband speeds, and the most advanced e-government in the world, Estonia—or E-stonia as Toomas Hendrik Ilves, its digitally savvy former president, calls it—is not only a "leader" in technology and "the next Silicon Valley"[1] but also the place where "stuff happens first."

I am at the Tehnopol, a campus of some four hundred start-up tech companies located in Tallinn Science Park on the edge of town, a short Uber ride away from the city's medieval Gothic walls. But rather than learning about the hottest new Estonian start-up, I am here to hear about the end of the world. I have come to listen to a speech on the existential risks of digital technology given by Jaan Tallinn, a cofounder of Estonia's two greatest digital success stories—the internet communications platform Skype and the peer-to-peer music-sharing website Kazaa. I am here to learn about what one of Europe's foremost technologists believes might be the darkest threat to the human species in its two-hundred-thousand-year history.

Tallinn is speaking at "Machine Learning Estonia," an informal conference about artificial intelligence. Outside, the January weather is as forbidding as the subject of his talk. It's not quite darkness at noon, but by midafternoon the sun has set and the outskirts of the Estonian capital are illuminated, if that's the right word, by the pale yellow of artificial lights. Sheets of fresh snow drifting into the city from the gray Baltic have blanketed the sprawling campus; I doubt that my networked driver could have found the boxy, nondescript building in the afternoon gloom without the help of the Google Maps app on his iPhone.

In contrast with the black, wintry weather outside, Jaan Tallinn—entrepreneur, educator, investor, philosopher, and civic leader as well as a technologist—is a man for all seasons. And so there is a stir in the audience, a murmur of anticipation when the diminutive, cropped-haired Estonian walks out onto the stage and—positioning himself between banners that proclaim, ACHIEVING THE FUTURE!—begins his speech. Dressed in blue jeans, a polka-dot T-shirt, and a hoodie with the logo FLEEP.IO emblazoned on its chest, the unsmiling computer scientist resembles a grown-up version of the baby-faced programmers in the audience. He is their future.

And he is speaking about the future too—not just the future of young Estonian computer scientists, but the future of us all. At most technology conferences, tech visionaries sell the promise of a bright future, but Tallinn's words are anything but utopian. His speech, "The AI Control Problem," which he gives in English, presents a dark new world in which self-conscious algorithms—computer code that thinks for itself—may "become anything without violating the laws of physics." In the future, we may no longer be in charge of our own creation, he suggests. Our technology might be developing a mind of its own, thereby excluding, disempowering, and enslaving us. The existential threat of self-conscious algorithms is very real, he says. They might be our final invention.

Technology, he warns, his voice rising slightly, is taking a "treacherous turn." Somebody in the audience nods. *Treachery*—that evil intent every Estonian knows by heart, given the tiny Baltic republic's bloody history of foreign meddling—might be at the bottom of it.

This warning of technology's treason, I suspect, has a more personal dimension too. As a cofounder of Skype and Kazaa, peer-to-peer websites designed to empower their users by giving them free telephone calls and enabling them to share music files, he is suggesting that digital technology might be *turning* on its inventors. The decentralized architecture built by technologists like Tallinn himself over the last fifty years could prove treasonous. What was designed to sit on the edge of the network is now hogging the center. What was created to enrich democracy is creating tyranny. That's the most serious treason for idealistic designers of the future such as Jaan Tallinn.

And it's treason with the most tragic of consequences. As Tallinn chooses to describe the tragedy, "The metaphor is as a parent seeing your children die."

Later, after his speech, when the audience has left, he meets with me. We sit alone, facing each other in the foyer of the Tehnopol building. Outside, the snow continues to fall on the city. It's perfectly quiet. Even the cleaning staff has gone home.

Tallinn in Tallinn. On the edge of Europe. Darkness on the edge of town.

He is weary, the rims of his eyes red with jet lag. Having returned to Estonia the previous day from Tokyo, he would fly out to New York the next morning. "Sometimes I joke that if the world gets destroyed by AI in the next ten years, the culprit is likely to be somebody I know," he tells me without smiling.

He is far from alone in fearing the taking over of the world by artificial intelligence, something that one leading American expert on AI describes as the "biggest event in human history." The world's richest man, Bill Gates; the world's most famous physicist, Stephen Hawking; Silicon Valley's most innovative entrepreneur, Elon Musk; and the Cambridge University cosmologist and author of *Our Final Century: Will the Human Race Survive the Twenty-First Century?* Martin Rees (Lord Rees), all share Tallinn's apocalyptic concerns. Not only do they all agree that this will be the biggest event in human history, they also fear it might be the *last*. In fact, this fear is so real that Tallinn—with Martin Rees and another illustrious Cambridge academic, Huw Price, the university's Bertrand Russell Professor of Philosophy—has cofounded a Cambridge research facility for studying the future: the Centre for the Study of Existential Risk.

But just ten years? I grimace. Is that all we have left? I ask.

Perhaps not in just a decade, but it could easily occur within fifty years, he says, echoing Lord Rees's fears that civilization won't survive the twenty-first century.

"We may no longer be in charge of the planet," Tallinn warns, pursing his lips grimly. Artificial intelligence, he tells me, really could "terminate" the human species.

A Nostalgia for the Future

Jaan Tallinn's apocalyptic vision of smart technology destroying humanity within fifty years is, of course, the worst-possible-case-scenario—probably better suited for a Hollywood movie script than an analytical nonfiction book like this. Yes, Tallinn and his fellow tech luminaries might be right to be concerned about the very long-term future of humanity in the face of increasingly intelligent technology. But, as Stanford University AI professor Andrew Ng acerbically noted, worrying about killer robots today is like fretting about overpopulation and pollution on Mars before we've even set foot on the planet.[2] No, the specter haunting us today is a much more familiar and down-to-earth one than the science-fictional scenario of humanity being enslaved by smart machines.

It is Karl Polanyi's "stark utopia"—that free self-adjusting market uncontrolled by either government regulation or the will of its citizens. It's the reappearance of the gaping inequalities, injustices, and mass addictions of the mid-nineteenth-century industrial economy. A return to a time when the will of regulators, consumers, educators, and citizens has been dwarfed by the seemingly rampant power of Polanyi's "utopian market economy."

There will be those who argue that the crises of the mid-nineteenth century—a "pestilence of hard liquor," the endemic child labor in "satanic mills," and those rivers of blood running through the streets of New York City—are of a more dystopian caliber than the birth pains of the digital revolution. But just as we now think about those times of mass exploitation and addiction with shame and horror, so—by the beginning of the twenty-second century—our great-grandchildren will look back at our own times with the same mixture of bemusement and disgust.

We are back in a world of rampant addiction, exploitation, lack of accountability, irresponsibility, and inequality. Today, the network effect of a mostly unregulated market has created tech companies of such astonishing power and wealth that they have become what the Oxford University historian Timothy Garton Ash calls "private superpowers."[3] Today we are all living under the big data spotlight of a surveillance economy in which we are incessantly watched by these corporate behemoths. Today we have returned to Polanyi's world of Two Nations, with America's nine richest tech billionaires being collectively wealthier than 1.8 billion of the world's poorest people. Today we have such an infestation of violent content in our digital media that it seems almost normal for online audiences of millions to watch revenge porn, live beheadings, and suicides. Today we have become so addicted to our networked devices that the eight-second average attention span of a human being is now shorter than that of a goldfish.[4]

So the theme of Jaan Tallinn's talk at the Tehnopol—the idea of treachery—remains extremely pertinent. Digital technology has, indeed, taken a *treacherous* turn. Not consciously, of course. As Ada Lovelace argued a hundred and fifty years ago, technology will probably never have a will of its own. But the internet revolution, which was supposed to empower us, is increasingly enslaving us. The web's decentralized architecture has become intensely centralized. What was created to enrich democracy is enabling a tyranny of virulent trolls and other antidemocratic forces.

"The internet is broken": thus conclude digital pioneers such as Twitter cofounder Evan Williams and Wikipedia cofounder Jimmy Wales.[5] Like Williams and Wales, more and more technologists are recognizing that today's networked transformation is writing us out of our own story. The internet might have been described as the "people's platform,"[6] these critics say, but in fact it has a people problem. Jaron Lanier, the inventor of virtual reality and Silicon Valley's most poignant thinker, even admits to a nostalgia for that halcyon time in the last century when technology did, indeed, put people first.

"I miss the future," Lanier confesses.[7]

He's far from alone. Even Tim Berners-Lee, the inventor of the World Wide Web, is nostalgic for the open, decentralized technological future he imagined he'd fathered in 1989. And so, at the 2016 "Decentralized Web Summit" in San Francisco, an event conducted in the same idealistic spirit as BlueYard Capital's "Encrypted and Decentralized" conference in Berlin, Berners-Lee spoke passionately about the state of the internet, particularly the emergence of vast digital monopolies and the pervasive culture of online surveillance. This summit, held in San Francisco's Inner Richmond district, near the Golden Gate Bridge, at the headquarters of the Internet Archive—the world's largest nonprofit digital library—captured the disenchantment with the current web that exists among many other leading technologists. Attended by such internet founding fathers as Berners-Lee and Vint Cerf—the inventor of the TCP/IP protocol that created the all-important "universal rulebook"[8] for global online communications, a code to enable the smooth running of the networked commons—the summit called for a return to the original sharing ideals of the web.

"We originally wanted three things from the internet—reliability, privacy, and fun," Brewster Kahle, the summit organizer and the founder of the Internet Archive, told me when I visited him at his funky offices in a defunct Christian Science church. We got the fun, he admitted. But the other stuff, privacy and reliability, he argued, hasn't been delivered. Privacy, in particular, remains a hugely important issue for Kahle. It was Edward Snowden, Kahle reminded me, who revealed that the British security apparatus was monitoring everyone who accessed the WikiLeaks site and then turning the names of American visitors over to the National Security Agency (NSA).

"That's frickin' frightening," Kahle, who was with Cerf and Berners-Lee a founding inductee into the Internet Hall of Fame, warns about this unaccountable online surveillance. "It shouldn't be a security decision every time you click on a link."

Part of the problem is that Silicon Valley's dominant business model—essentially the commercial appropriation of users' data—is profoundly flawed. Facebook, Google, YouTube, Instagram, Snapchat, WhatsApp, and most of the other dominant internet businesses give away their products at no charge and then make all their money by selling increasingly personalized and intelligent advertising around these free products. Of the $89.46 billion in Google's 2016 annual revenue, for example, $79.38 billion was generated by advertising revenue. So what Kahle calls "unaccountable online surveillance" is business as usual in Silicon Valley. "The primary business model of the internet is built on mass surveillance," concludes Bruce Schneier, one of America's leading computer security experts.[9]

The scale of this data economy is mind-boggling. Three and a half billion internet users around the world create 2.5 quintillion bytes of data each day. In every minute of every day of 2016, we made 2.4 million Google searches, watched 2.78 million videos, entered 701,389 Facebook log-ins, added 36,194 new posts to Instagram, and exchanged 2.8 million messages on WhatsApp.[10] All this personal data has become the most valuable commodity of the networked age, the so-called new oil[11] of our networked economy—as everyone from European politicians to Silicon Valley venture capitalists to the CEO of IBM has put it—endowing Big Tech with the wealth of the gods.

Tim Berners-Lee shares Kahle's disappointment with recent digital history. "The Internet was designed to be decentralized so everybody could participate," Berners-Lee told the summit attendees about a digital architecture he helped design. Instead, he said, "personal data has been locked up" in what he called "silos"—centralized big data companies like Google, Amazon, Facebook, and LinkedIn.[12] And so "the problem," he warned, "is the dominance of one search engine, one big social network, one Twitter for microblogging."

"We don't have a technology problem," Berners-Lee concluded about the new operating system that he inadvertently helped build; "we have a social problem."[13]

But it's more than just a social problem. As with the impact of the industrial revolution on nineteenth-century life, today's digital revolution represents a civilizational problem that is disrupting our politics, economics, culture, and society. "As the Silicon Valley giants prosper," warns the *Financial Times*'s Rana Foroohar, "everyone else is falling behind."[14] Polanyi's utopian market economy has thus reappeared in a digital form. Once again, things are radically out of sync. And, once again, it's a problem that can be addressed only by the combined work of regulators, educators, innovators, consumers, and citizens.

Private Superpowers:
The Four Horsemen of the Apocalypse

In 2015, I published my third book, *The Internet Is Not the Answer*,[15] a work that addressed the skewed distribution of power and wealth in the network age. The tragedy of today's digital revolution, I argued, is that the ideals of digital pioneers like Norbert Wiener, Tim Berners-Lee, Brewster Kahle, and Jimmy Wales—democracy, equality, enlightenment, freedom, universality, transparency, accountability, above all public space—have not, so far at least, been realized. Instead of Berners-Lee's public World Wide Web, the online revolution has been appropriated by Garton Ash's private Silicon Valley superpowers.

I suggested that digital technology might, indeed, be *turning* on its inventors. What was designed to be a decentralized, egalitarian, and enlightened revolution, a kind of digital commonwealth, I warned, is turning out to be disturbingly centralized, unequal, and creepy. This digital spin on a utopian market economy is creating a "winner-take-all network,"[16] I argued, and we, the people, we are all the losers in it.

Take, for example, the impact of the digital revolution on the mapmaking industry. The goal of this book, you'll remember, is to create a map of the future that places humans at its center. But the winner-take-all nature of the new economy has put money, rather than people, at the heart of today's mapmaking industry. "We are on the brink of a new geography," explains Jerry Brotton, the Professor of Renaissance Studies at the University of London and an acclaimed historian of maps. The new digital mapmaking industry, Brotton warns, "risks being driven as never before by a single imperative: the accumulation of financial profit through the monopolization of quantifiable information."[17] And the *single* company driving this imperative is Google which, in May 2016, announced plans to add location-based advertising to its mobile map products, thereby making the tracking of our location data ever more valuable.[18]

This is particularly troubling, Brotton argues, because the Silicon Valley big data company—which owns Android, the mobile operating platform currently used by over 86 percent of all smartphone owners in the world[19]—does not "disclose the specific details of its code." And so, "for the first time in recorded history," Brotton observes about an industry that began more than two millennia ago in Alexandrian Egypt with Ptolemy's *Geography* and flowered during the original Age of Discovery with the global maps of Renaissance cartographers like Diogo Ribeiro and Gerard Mercator, "a world view is being constructed according to information which is not publicly and freely available."[20] The geography of our new world, then, will be a privatized one—a "walled garden" to borrow a Silicon Valley euphemism for monopolies—owned and operated by what Tim Berners-Lee calls vast data "silos." The geography of the commons is being replaced by the geography of the proprietary algorithm; it is a new geography in which we, the people, will be simultaneously excluded and tracked. Geography is in danger of becoming entirely privatized. *Memento mori*, indeed.

Or take the dramatic changes over the last quarter century in media and entertainment, the first industries to be radically affected by the digital revolution. According to Jonathan Taplin, a longtime movie producer and music promoter, there has been a "massive reallocation" of wealth between "the creators of talent" and "the owners of monopoly platforms" like Facebook, YouTube, or Google. Taplin—who managed Bob Dylan's early tours and produced movies for Martin Scorsese, Wim Wenders, and Gus Van Sant—argues that between 2014 and 2015, some $50 billion per year flowed from the traditional creative industries to the new monopolists of Silicon Valley.[21] In contrast with the traditional entertainment businesses, companies like YouTube don't actively invest in talent, curate content, or sell culture. They are simply the places people go on the internet to listen to free content. In music, for example, YouTube mostly makes its money by selling advertising rather than investing in talent, and it pays the artists and labels a pittance—less than one-tenth of a cent per song played.[22] It is a rentier—like one of those unproductive sixteenth-century English landlords parodied in More's *Utopia*. It lives off its increasingly monopolistic position as the winner-take-all player in online entertainment.

Then there's the impact of the digital revolution on publishing and, by extension, on politics. "Our news eco-system has changed more dramatically in the past five years than perhaps at any time in the past five hundred years," says Emily Bell, the director of the Tow Center for Digital Journalism at Columbia University's Journalism School.[23] Digital technology, she warns, has actually put the future of publishing "into the hands of a few, who now control the destiny of many."[24] Social media, Bell says, have "swallowed journalism" with what she dubs—with a medieval symbolism that might have amused Erasmus of Rotterdam—the "Four Horsemen of the Apocalypse," Google, Facebook, Apple, and Amazon, engaged in a "prolonged and torrid war" for our attention. These private superpowers are thus becoming our "new speech governors," usurping the traditional role of government in determining what can and can't be published.[25] Meanwhile, online publishers—the actual creators of content and, one would assume, economic value—remain mired in crisis, with 85 percent of all online advertising revenue going in the first quarter of 2016 to just two of Tim Berners-Lee's centralized data silos: Facebook and Google.[26]

The monopolization of media isn't just a problem for publishers. With Facebook as our new front page on the world, we are simply being refed our own biases by networked software owned by a $350 billion data company that resolutely refuses to acknowledge itself as a media company because that would require it to employ armies of real people as curators. It would also make Facebook legally liable for the advertising that appears on its network. So what we see and read on social media, therefore, is what we want to see and read. No wonder everything now seems so *inevitable* to so many people. This echo chamber effect, the so-called filter bubble,[27] has created a hall of mirrors, a "post-truth" media landscape dominated by fake news and other forms of online propaganda. Thus the disturbing success of Trump, Brexit, and the alt-right movement; thus the virulence of Putin's troll factories, networked ISIS recruiters, and the other mostly anonymous racists, misogynists, and bullies sowing digital hatred and violence.

The 2017 Edelman Trust Barometer captures this deeply troubling connection between the global implosion of trust and the rise of our echo chamber media. The respondents to the Edelman report favored search engines (59 percent) over human editors (41 percent), and the Barometer found that they were nearly four times more likely to disregard online information that supports a position in which they don't believe. "The lack of trust in media has given rise to the fake news phenomenon and politicians speaking directly to the masses," warns Richard Edelman.[28]

As I predicted in my 2007 book, *The Cult of the Amateur: How Today's Internet Is Killing Our Culture*, the ultimate victim of this supposed "democratization" of media is the idea of truth itself. Without gatekeepers, fact-checkers, or editors able to verify the truth of a newspaper article or television news report, your fake news is as "truthful" as mine.

America now even has a democratically elected president—as every schoolchild knows, the most powerful person in the world (one of the few remaining "truths" on which everyone, outside Moscow, agrees)—who is a product and a dispenser of this corrosive fake news culture. And he's made it abundantly clear that his real enemy is what's left of objective, curated media. Even some of Silicon Valley's most powerful people are finally beginning to recognize the seriousness of this problem. Apple CEO Tim Cook, whose company's business model, unlike those of Google and Facebook, doesn't depend on the sale of its users' data, argues that "fake news is killing people's minds."[29] Cook believes that the problem requires public action. Like the campaigns to educate people about our environmental crisis, he says, it needs a concerted public effort to educate people on the dangers of fake news.

But it's not only fake online news that is killing people's minds. Just as much of the industrially produced food of the mid-twentieth century was designed to be addictive, so too many of today's digital products have been created to hook us on them. There are even bestselling books by Silicon Valley "marketers," with seductive titles like *Hooked*, that are essentially manuals explaining how to build habit-forming products.[30]

One high-profile addict is the Anglo-American blogger and polemicist Andrew Sullivan, a "very early adopter" of what he calls "living-in-the-web,"[31] who has admitted attending a meditation retreat center in Massachusetts to kick his digital habit. "I used to be a human being," Sullivan confessed about the consequences of his internet addiction. According to the New York University psychologist Adam Alter, there are many millions of others like Sullivan with what Alter calls a "behavioral addiction" to our smartphones and our email.

"In the 1960s, we swam through waters with only a few hooks: cigarettes, alcohol and drugs that were expensive and generally inaccessible," Alter argues. "In the 2010s, those same waters are littered with hooks. There's the Facebook hook. The Instagram hook. The porn hook. The email hook. The online shipping hook. And so on. The list is long—far longer than it's ever been in human history, and we're only just learning the power of these hooks."[32]

Just as Tim Cook is calling for public action against fake news, so other Silicon Valley technologists are calling for a similar response against what the San Francisco–based pediatric endocrinologist Robert Lustig calls "the hacking of the American mind" by Alter's digital hooks. According to the *Financial Times*' Izabella Kaminska, this hacking is directly undermining our happiness. "Despite all the technology that connects us, much of it there supposedly to make our lives easier and better," Kaminska warns about the impact of our digital hooks, "people have never been more depressed."[33]

A former "Design Ethicist" at Google, Tristan Harris, has gone as far as to argue that software developers should sign a code of conduct, a kind of Hippocratic oath, promising to create products that treat their users with respect. Harris, who is the founder of the nonprofit Time Well Spent and was described by the *Atlantic* as "the closest thing Silicon Valley has to a conscience," believes that there need to be "new ratings, new criteria, new design standards, new certification standards" to ensure against addictive products.[34] When I interviewed Harris for my TechCrunchTV show, he told me that the three dominant digital platforms of our networked age—Apple, Google, and Facebook—have been explicitly designed to hook our attention.

"We all live in a city called the attention economy," Harris argued, citing the work of the urban planning pioneer Jane Jacobs and suggesting that the challenges of software developers are similar to those of urban planners. And if this twenty-first-century city is to be habitable, he said, then these platform owners need to take responsibility for the impact of their products on their users.

Can the Internet Save the World?

In *The Internet Is Not the Answer*, I argued that the digital revolution, combined with the network effect of a winner-take-all economy, has created a Big Tech concentration of power that dwarfs Big Banks, Big Oil, or Big Pharma in its size and significance. It's a much more lavish meal than just Facebook *swallowing* journalism, I suggested. Corporations like Google and Facebook were once known, rather quaintly, as "internet" companies. But today they are rapidly becoming artificial intelligence companies and self-driving car companies and virtual reality companies. "Software is eating the world," is how Marc Andreessen—the cofounder in 1994 of the first internet browser company Netscape and the original boy king of Silicon Valley—describes the way in which networked technology is feeding on almost everything and everyone else. Silicon Valley isn't just the new Wall Street—it's actually wealthier and more powerful than the old Wall Street, outspending the financial industry by two to one in Washington, DC, lobbying during the Obama administration's eight years.[35]

Yes, we've seen this before. It's the reappearance of the same utopian market economy creating the same radical inequalities between rich and poor that have appeared many times before in history—from the enclosures of Thomas More's sixteenth-century English countryside that resulted in sheep "devouring" men to the overcrowded industrial tenements of nineteenth-century New York City. But this time the numbers are even more astonishing. The five most valuable companies on the Fortune 500 list—the Four Horsemen of the Apocalypse plus Microsoft—are all West Coast American tech corporations. Collectively, these five winner-take-all technology companies, the "frightful five," according to the *New York Times*' technology writer Farhad Manjoo, which together employ a little over half a million people, are worth around $2.3 trillion. If they were a country, this would rank them as the seventh-largest economy in the world, a couple of hundred billion dollars ahead of the annual GDP of India, with its more than 1.2 billion inhabitants.[36]

The personal wealth of the founders of these companies is equally astounding. The nine richest tech billionaires in Silicon Valley have a collective wealth that is more than that of 1.8 billion of the world's poorest people, a quarter of the world's entire population. "It is obscene for so much wealth to be held in the hands of so few when one in ten people survive[s] on less than two dollars a day," noted Oxfam's executive director, referring to the surreal chasm between a handful of Silicon Valley entrepreneurs and the rest of the world.[37] This obscenity has a more local context too. In impoverished East Palo Alto, the neighboring town to Palo Alto, the home of Stanford University and the epicenter of Silicon Valley, more than one-third of the schoolchildren are homeless.[38]

And part of the problem is that the economic gains of smart technology aren't trickling down to the rest of us. Silicon Valley's abundance of capital might even be compounding scarcity elsewhere. The people problem, if anything, might be getting worse. As the Columbia University economist Jeffrey Sachs told me about our Two Nations world, "something is going on in national income distribution which is connected to the smart machine." Sachs is concerned that technology that improves technology might "be making people worse off." There is "truth to the idea that machines can take jobs" and even "lower our quality of life," he explained to me. It's the shift in the distribution of wealth from labor to capital that really worries the economist. "And it's certainly not Luddite to have these forebodings," he insisted.

Sachs acknowledges that the same debate has been raging for the past two hundred years, ever since James Watt invented the steam engine and a combination of a mechanized textile industry and the development of railroads created the first industrial civilization. For two hundred years, Sachs admits, we've been debating whether ever-more-powerful machines would enslave or empower us. But now, he insists, the question of the impact of smart technology on jobs is "becoming urgent." What concerns him is that history is repeating itself in the sense that—as in the first blush of the industrial revolution—technology might be enlarging the economic pie but is failing to establish a "new prosperity shared by all."[39]

Sachs's concerns are increasingly shared by Silicon Valley notables as well as by politicians traditionally sympathetic to the tech industry. Gavin Newsom, California's current lieutenant governor, has gone as far as to say that it's the job of Silicon Valley technologists to "exercise" their "moral authority" in fighting income inequality and job losses, which he describes as a "code red, a firehose, a tsunami" that is rapidly approaching us. "The plumbing of the world is radically changing," Newsom said in 2017, rendering his own rousing version of More's Law to graduating computer science students at the University of California, Berkeley; "your job is to exercise your moral authority. It is to do the kinds of things that can't be downloaded."[40]

A World Bank study, *World Development Report 2016: Digital Dividends*, also suggests that the digital revolution may be widening inequality and hollowing out middle-class jobs. This report shows that the supposed dividends of "higher growth, more jobs, and public services" from the digital revolution have "fallen short of expectations."[41] The research project, cochaired by the then–Estonian president, Toomas Hendrik Ilves, and the World Bank's chief economist Kaushik Basu, found a global walled garden in which 60 percent of people in the world remain entirely excluded from the digital economy. Rapid digital expansion, the report's authors warned, "has been skewed towards the wealthy, skilled, and influential around the world, who are better positioned to take advantage of the new technologies."[42] It's the winner-take-all network, a Two-Nation digital universe writ large.

"Can the internet save the world?" asks the *New York Times* about this 2016 World Bank report.[43]

The answer, for the moment at least, is no.

But how *can* the internet—as what Gavin Newsom calls "the plumbing" for our new twenty-first-century operating system—actually help save the world?

As I suggested at the beginning of this chapter, the future sometimes appears in the unlikeliest of places. And the European country that is pioneering a better digital society by combining free market innovation with equally creative regulatory and educational reforms in order to put people back in the center of the digital map is the little Baltic republic of Estonia—or E-stonia, as Toomas Hendrik Ilves, its digitally savvy former president, calls it.

So let's return to Estonia. Where stuff happens first.

Chapter Four

UTOPIA: A CASE STUDY (BOOK ONE)

A Country in the Cloud

Fortunately, not everyone in Estonia is as miserable as Jaan Tallinn. Indeed, the day after I hear his lecture about the end of the world, I spend the day in Tallinn meeting some of the remarkable policy makers who are cheerfully transforming the little Baltic republic from a backward Soviet colony into the next Silicon Valley.

Figuring out how to get around the Estonian capital is easy—if you have a map, that is, or at least some mental geography of the journey. We may have Waze, Google Maps, and the other GPS-powered travel apps on our smartphones, but we still need to physically transport ourselves by train, cab, or foot from one place to another to reach our destination. There are no superintelligent teleportation machines on the technological horizon—not even in high-tech Estonia. Geography, as one world traveler reminds us, continues to matter.[1] And the "first law of geography," another leading geographer explains, is that "everything is related to everything else."[2]

There are few places in the world where geography has mattered in the past more than in Estonia, the country with the ill fortune to share a border with Russia and a sea with Denmark, Sweden, and Germany—regional powers that have all, over the last half millennium, been rather too *related* to the little Baltic republic. And yet in spite of Estonia's unfortunate geography, or perhaps because of it, this land of just twenty-eight thousand square miles is today— with South Korea, Israel, Singapore, and its Scandinavian neighbors—one of the most wired and innovative countries on the planet. The government and people of Estonia are trying to invent the ideal information society. They are figuring out how to live well in cyberspace.

Estonia is, indeed, an e-society with an e-state and an e-government that is reimagining cyberspace as civic space, from its e-residency program to its focus on the internet in schools to the idea of government as a service. "Estonians embrace life in the digital world," the *New York Times* tells us about a society that, it says, "lives first and foremost online."[3] The Baltic republic is one of the "most internet-dependent countries in the world," explains the BBC.[4] Estonia, adds the *Atlantic*, has the "world's most tech-savvy government."[5]

No, geography isn't always destiny—not, at least, in Estonia. The Canadian new media guru Douglas Coupland suggests that in today's networked economy "your brand is your border." But Estonia's remarkable achievement has been to build a digital brand that is significantly more expansive than its analog borders. That's because "we are running a country in the cloud," said Estonia's chief technology officer, Taavi Kotka, when I visited him at his little office just off Tallinn's Town Hall Square, the eleventh-century Gothic heart of the old town.

Indeed, as the first country in the world to offer "e-residency"—an electronic passport that offers any small businessperson the right to use legitimate Estonian legal or accounting online services and digital technologies—Estonia is even trying to disrupt the age-old intimacy between physical territory and citizenship. This e-residency program underwrites online identity by establishing fingerprints, biometrics, and a private key on a chip.

The goal of the program, its twenty-eight-year-old director, Kaspar Korjus, tells me, is to have ten million e-residents citizens by 2025—that's almost eight times the number of Estonia's current population of 1.3 million. Korjus wants to create what he calls a "trust economy" for businesspeople around the world. The e-residency program is a well-lit antithesis to the dark web—the digital hell infested with drug and arms dealers, pedophiles, and other criminals. "We want to be the Switzerland of the digital world," Korjus explains. Taavi Kotka is more ambitious. We are becoming "the Matrix," the Estonian CTO tells me without smiling.

VISIT ESTONIA BEFORE ESTONIA VISITS YOU, as one T-shirt on sale at the airport said.

The Estonian government puts it slightly more circumspectly. "Moving toward the idea of a country without borders" is how it describes this audacious e-residency scheme.

It's an example of "outsourcing government," Kotka explains. Given its unforgiving climate and inconvenient geography, Estonia has always struggled to find physical residents, he explains. So e-residency creates a platform for a new global kind of citizenship. Not only, therefore, is Estonia running a government in the cloud, it is also trying to create a country in the cloud—a twenty-first-century distributed community of people united by networked services rather than by geography. It's Kotka's matrix—a country operating in the infinite realm of cyberspace.

Not surprisingly, innovation tourists like myself flock to see this country without borders, where stuff happens first. The Baltic state is, according to Alec Ross—Hilary Clinton's digital guru while she was secretary of state, and a keen student of Estonian innovation—the "little country that could." With an "entire economy [that] seems to be an e-economy," Estonia is, Ross promises, a country of the future, so naturally rich in digital innovation that its citizens are producing start-ups—like Jaan Tallinn's Skype and Kazaa—that "make Silicon Valley green with envy."[6] Ross contrasts Estonia with another prisoner of geography, nearby Belarus, which has totally stagnated as a parochial command economy. "While Estonia opened up," Ross concludes, "Belarus closed off."[7]

The first law of geography might be that everything is related to everything else, but in 2017 the first law of Estonia is that practically everything and everyone is connected to the internet. And so 91.4 percent of Estonia's 1.3 million citizens are internet users; 87.9 percent of its households have computers; 86.7 percent of Estonians have access to broadband; and 88.4 percent of them use it regularly.[8] In neighboring Latvia, by contrast, only 76 percent of its citizens are internet users, while in Russia, Estonia's former colonial ruler, that number is just 71 percent. Yet digital access alone, as then-president Ilves warned in the World Bank report, is an insufficient strategy for realizing transformational change. To build a genuinely networked society, he insisted, a country needs a mix of regulation, legislation, innovation, and education.

"The experience of Estonia shows that internet access alone is not enough to reap the benefits of digital development, as most work has to be done at the level of state governance, legislation, and education," Ilves told a Washington, DC, audience at the launch of the World Bank report in January 2016.[9]

The education system has played a particularly important role in improving the skills of people in Estonia, Ilves explained. This, indeed, was the first step in the E-stonian revolution. By the late 1990s a government-backed investment fund was paying for access of all schools to the internet, as well as teaching computer programming skills to kids as young as seven. "It's like literacy," one software engineer told me, describing the way essential programming skills are now viewed within schools.

The Estonian educational system is also being redesigned to make people more responsible citizens. As Kristel Rillo—who runs e-services at the Ministry of Education—explains, Estonian schools now have obligatory programs called "digital competence." The country is even planning a national test for digital competency in five areas, including the correct use of netiquette. Education is "two steps ahead of the labor market," Rillo tells me, adding that kids are turning out to be "two steps ahead of middle-aged workers in learning how to become digital citizens."

Trust, Trust, Trust

But the most intriguing step in Estonia's digital development has taken place outside the classroom. The key to the Estonian digital revolution is an identity card system that puts digital identity and trust at the heart of a new social contract. The mandatory electronic ID card, currently used by more than 95 percent of Estonians, gives everyone a secure online identity and offers a platform for digital citizenship featuring more than four thousand online services, including the storage of health and police records, paying taxes, and voting.

This online ID system is an attempt, explains Andres Kütt, the chief architect of the Estonian Information System Authority, to "redefine the nature of the country" by getting rid of bureaucracy and reinventing government as what he calls "a service." As one of the principal architects of this new system, Kütt aims to integrate everyone's data into a single, easy-to-navigate information portal. Kütt, a recent MIT graduate and former Skype employee, says that Estonia wants to smash bureaucratic silos and distribute power down to the citizens so that government comes to them rather than their having to go to the government.

I meet Kütt in his government office on the top floor of a run-down building hidden behind a strip mall. Dressed in a green knitted sweater, he's a small man with a wispy beard and an unashamedly hobbit-like enthusiasm for the ambitious technology he is building. "The old model is broken. We are changing the concept of citizenship," he says by way of explaining what he calls "government-as-a-service." "This technology creates trust. It's transparent. All agencies can access this data, but citizens have the right to know if their data has been accessed. In the old world, citizens were dependent on government; in Estonia, we are trying to make government dependent on citizens."

The ID system Kütt is designing is supposed to be the reverse of Orwell's Big Brother. In Estonia, citizens are being empowered to watch the operations of government. And although the government can look at people's data, it must notify them when it does so. Kütt gives me a personal anecdote of how the system works. He had driven to a lecture in Tallinn to demonstrate the ID system. When he looked at his data, he saw that a police officer had accessed his information thirty minutes earlier. Following up in his online records, he found that an unmarked police car had followed his car on the road to Tallinn because his license plate was dirty. The police accessed his records, checked his driver's license, and decided not to stop him. The point of this story, Kütt reminds me, is to stress the accountability of government in this new citizen database system. Nothing can be done secretly, he insists. It's a transparent system designed to protect individual rights and compound the trust between citizens and their government.

Everyone I spoke to in Estonia—from start-up entrepreneurs to policy makers to technologists to government ministers—agreed with Kütt that the most important aspect of the ID system is the creation of trust. It's like that old real estate cliché about the three most important qualities of a house: *location, location, location*. Likewise, the three most important things about the Estonian ID system are *trust, trust, and trust*. Given what the 2017 Edelman Trust Barometer described as "the implosion of trust" in the world today, this makes the Estonian experiment relevant for all of us.

Estonians "trust" their government, Merle Maigre, the country's security policy chief, promises me. "Everything relies on trust in digital society," explains the country's CTO, Taavi Kotka. Estonia, adds Kaspar Korjus, the director of the e-residency program, is developing what he calls "the trust economy." And the government, Korjus explains, is the only institution that can credibly underwrite trust for the whole community.

Sten Tamkivi, a start-up internet entrepreneur who formed Estonia's first digital advertising company in 1996 while he was still in high school and sold it later to the global agency DDB Worldwide, describes the ID system as a "trust mechanism." Tamkivi tells me that while it might be mandatory, "your data is yours." You get to see who looks at your records, he says, and, reiterating what everyone else told me, only the government can access personal data in a system that has been designed to be so transparent that individuals are notified whenever the authorities access it.

Ain Aaviksoo, deputy secretary-general for e-services and innovation at the Ministry of Social Affairs and part of the team that built Estonia's first health portal, also sees this new system as compounding trust. "The Estonian people trust the system because they haven't seen misuse. This system gives people the ability to determine their own privacy, but they have to take responsibility," he tells me.

According to Siim Sikkut, a Princeton-educated technologist who is now an advisor to the Estonian president, this national ID system ensures that you really are who you claim to be. Like the e-residency scheme, it "authorizes identity," Sikkut explains. And the only person who has full access to your data is you. "After all, if you can't trust your government," Sikkut asks rhetorically, "then who, exactly, can you trust?"

Estonian trust in government is, in fact, much higher than the EU average. A 2014 Eurobarometer study found that 51 percent of Estonians trust their government, in contrast with the EU average of 29 percent. But while Estonians' trust in their government is high, the study found that only 13 percent of Estonians trusted their political parties.[10] Given Estonia's commitment to reinventing government as an online service, this sharp disparity between people's trust for government and their trust for political parties can be explained by the success of their digital reforms.

"Estonian people don't trust government, but they do trust e-government," explains Linnar Viik, another of the architects of the ID card infrastructure and a serial tech entrepreneur dubbed by the press "Estonia's Mr. Internet."[11]

I ask Viik to explain the technology behind the ID system that can guarantee trust. It is, Estonia's Mr. Internet tells me, "public infrastructure built with asymmetric, distributed technology" containing "digital signature architecture with a time stamp." What this means, in simpler language, is that the information entered into the system can't be altered or even looked at unless the owner of the data is alerted. The medical or financial or criminal records in the system are, therefore, guaranteed to be trustworthy. They can't be secretly tampered with or spied upon. Viik describes this as "blockchain before blockchain." Blockchain, by the way, is new technology that enables the creation of a public database that can't be tampered with or altered. According to the Canadian futurists Don and Alex Tapscott, this blockchain technology, which they call the "trust protocol," could be the most significant technological development since the invention of the internet.[12] So what Viik means by his "blockchain before blockchain" comment is that while the Estonian ID system doesn't contain formal blockchain technology, it nonetheless has a similar impact: creating what the *Economist*, in describing blockchain, called a "trust machine . . . the great chain of being sure about things."[13]

As always with new *new* things, the technological details of this process are less interesting than its political and economic consequences. And there are some extremely significant potential consequences of the Estonian government's entry into the data business. According to Viik, one result might be a new type of rivalry between sovereign governments and Silicon Valley's private superpowers.

"Governments are realizing that they're losing the digital identities of their citizens to American companies like Google, Facebook, Amazon, and Apple," Viik explains. "And they are waking up to the realization that they have a responsibility to protect the privacy of these citizens."

It's personal data, of course, that has made these private superpowers so wealthy and powerful. And though the ID system of course doesn't stop Estonians from using Facebook or Google, the database is designed as a rival kind of ecosystem, a secure public alternative designed to benefit citizens rather than private corporations.

Viik, like Betaworks CEO John Borthwick, argues that one of today's great challenges is to reinvent the relevance of government in the new digital world. And it's here, he believes, that the longer-term significance of the ID system may lie. "The government's role is to protect the privacy of its citizens," Viik says about this Estonian government policy of underwriting personal data. "It's an extension of public infrastructure—the twenty-first-century version of the welfare state."

To some readers, particularly those, like Edward Snowden, who cherish their privacy, this ID system, with its radical transparency, might sound a little dystopian. But one of the unavoidable consequences of the digital revolution is the massive explosion of personal data on the network. Like it or not, this data is only going to grow exponentially with the development of smart homes, smart cars, smart cities, and, above all, all the other smart objects driving the internet of things. We don't have a choice about any of this. But what we do have a choice about is the amount of transparency we demand of the governments or corporations that have access to our personal data. That's why Viik's attempt to create a blockchain-like transparency within the Estonian government's database of information about its citizens is so important. It may not be an ideal solution but, outside of Utopia, this Estonian model is probably about as good as we are going to get.

Speaking of Utopia, perhaps it should come as no surprise that in Thomas More's little sixteenth-century book, the island of Utopia was organized on transparent principles similar to those in twenty-first-century Estonia. In *Utopia*, there is "nothing private anywhere," with people even being able to go into somebody else's house and snoop around if they felt like it.[14] Things were indeed so open in Utopia that there was a custom that husbands and wives, before marriage, would be required to inspect each other naked to make sure they knew exactly what they were getting in the transaction.[15]

And so, five hundred years after Thomas More wrote about that transparent no-place, founded by King Utopus, we now have E-topia. But unlike More's sixteenth-century fantasy, twenty-first-century Estonia can be found on a map. It even boasts a real-life version of Utopus—the tech-savvy Toomas Hendrik Ilves, the Estonian president between 2006 and 2016.

E-Topus

The Presidential Palace in Kadriorg Park couldn't be more different from the Tallinn Science Park building where I spoke with Jaan Tallinn at the "Machine Learning Estonia" meet-up. This palace is an ornate early-twentieth-century building situated in a tree-lined private park near the cobbled streets of the old town. It is designed to resemble the early-eighteenth-century Kadriorg Palace, a frilly baroque building that was commissioned by Peter the Great for his wife, Catherine (Kadriorg means "Catherine's Valley"

in Estonian), after the successful Russian siege of Tallinn
in 1710.

The difference between Estonia's two most famous
technocrats—Toomas Hendrik Ilves and Jaan Tallinn—is
equally dramatic. In contrast with the reticent and diminutive Tallinn, President Ilves, the country's most powerful
legislator, is a robust man with a media-ready voice and the
colorful manner of an intellectual showman. He's wearing
the dapper uniform of blazer, dress shirt, and bow tie to
match. And while Tallinn keeps his private life intensely
private, Ilves's hundred thousand Twitter followers can keep
track of everything he does, including the details of his three
high-profile marriages and his various offspring.

The country without borders, it seems, has a president
without borders. Sitting across the table from me in one of
the palace's formal dining rooms, Ilves tells me the story
of his life. Technology provides the thread to his narrative.
Born in 1953 in Stockholm to Estonian refugees who had
fled the postwar Soviet occupation, Ilves grew up in New
Jersey, where he was both an academic prodigy and an early
computer geek. He boasts that he learned to program as a
thirteen-year-old and that he owned an Apple IIE personal
computer, the immediate predecessor to the Macintosh. He
earned undergraduate and graduate degrees in psychology
from Columbia University and the University of Pennsylvania and then, as a radio journalist, ran the Estonian desk
at the Munich-based Radio Free Europe. When Estonia—
which, he reminds me, was "very poor"—became independent in August 1991, he'd asked himself rhetorically, "What
do we have here?"

"Good math skills," he'd answered. "That's what we *have.*" Under the Soviet occupation, Ilves explains, little Estonia had had the good fortune of being used as a kind of research and development lab for advanced technology. And so, unlike those in other East European countries colonized by the Soviets, Estonia's universities hadn't been ransacked; and its population—including gifted young technologists like the nineteen-year-old Jaan Tallinn—remained among the best educated in Eastern Europe. Estonia's future, therefore, Ilves concluded, would be "high tech."

Much of this was inspired, he tells me, by what he calls a "backward reading" of Jeremy Rifkin's influential book *The End of Work*, which predicted the decline of the workforce in a postindustrial age.[16] What Ilves took, however, from Rifkin's thesis is the counterintuitive idea that small countries like Estonia could actually benefit from the great transformation from an industrial to an information economy. Size would matter in the future too, he figured, but while the twentieth century naturally advantaged industrial leviathans like the Soviet Union, with their massive economies of scale and millions of industrial laborers, the networked age could favor small start-up nations like Estonia, with its small, highly skilled workforce that could speedily pivot around new technologies. The future, Ilves correctly predicted, would belong to countries nimble enough to perpetually reinvent themselves in our age of constant change.

It would also belong to nimble people, like Ilves himself. After the country gained independence in 1991, Ilves successfully reinvented himself, pivoting from radio journalism to Estonian politics. In 1993 he was appointed Estonian ambassador to the United States. From 1996 to 1998 and 1999 to 2002 he was the Estonian minister of foreign affairs. In 2004 he became a member of the European Parliament. Then, in 2006, he was elected by Estonia's parliament as the country's fourth president, a five-year post he has held twice.

As a public servant over the last quarter century, he has led the transformation of Estonia from a forgotten northwestern province of the Soviet Union into a hub of high-tech innovation. Ilves is a personification—albeit a rather noisy and self-confident one—of More's Law. He helped pioneer the computerization of schools as well as the establishment of internet centers and public WiFi all around Estonia. He has led the investment in such infrastructure as the Tehnopol campus, with more than four hundred tech start-ups in Tallinn Science Park. He has helped with the digitalization of Estonian public records and books so that, in case of another Russian occupation (not inconceivable given that relations with a revanchist Moscow are increasingly tense), this data will remain safe. One of his most cherished accomplishments is the e-residency program, which offers non-Estonians digital citizenship. And through his high-profile work with such international organizations as the World Bank and the World Economic Forum, the colorful Ilves has certainly made sure that little Estonia appears to the outside world as a country without boundaries.

Ilves's major legacy lies, however, in his pioneering work with what he calls "digital identity and trust."

"The role of the sovereign in a digital society," he explains to me, "is to guarantee identity."

I remember Edward Snowden's speech at the Alte Teppichfabrik. "What does it mean when we are all transparent and have no secrets anymore?" Snowden had asked the Berlin audience. The answer, he suggested, was the undermining of individual freedom and of the other verities of the self, as articulated by liberals like Samuel Warren and Louis Brandeis. So, for Snowden, the role of the sovereign is very simple. It's to keep out of our online business, to allow us to maintain our individual autonomy, to leave us alone. Ilves's notion of "guaranteeing identity" sounds much too intrusive for a privacy advocate like Edward Snowden. So I raise the Snowden question with Ilves: Is Snowden a heroic leaker or traitorous hacker?

Ilves is much too wily a political fox to fall into the trap of defining Snowden as either. What he does tell me, however, is that the NSA contractor, now exiled somewhere in Russia, was wrong to be so obsessed with our privacy from government agencies. Snowden, he says, created a "paranoid shit storm" over his revelations about the NSA, even though most of these revelations were, in Ilves's opinion, either completely misunderstood or inaccurate. Rather than reading "bohemian poets' emails to their girlfriends," he says, the NSA, in its legitimate search for terrorists, was simply using the metadata to "look at who is connecting to who.

"Our obsession with privacy is misguided," Ilves insists, wagging his finger at me so insistently from across the table that I worry his bow tie might spin off. "The real issue is data integrity."

It's not that Ilves is completely dismissing the significance of privacy or diminishing its importance as a component of individual freedom. But the *real issue* for government in an internet world, where data about everything and everyone is readily accessible on Google, he says, is refereeing a system where that data is authentic. He is saying that there's nothing more important than *data integrity* in a digital twenty-first century, where everything—including ourselves—is turned into information. And so the role of government is to create a trustworthy informational exchange system that is secure. If data is, indeed, the new currency of the networked age, then it requires what he calls a "sovereign guarantee." Like currency, it has value only if it has this official stamp of authenticity.

"Somebody knowing my blood type isn't a big deal," he explains. "But if they could change the data on my blood type—that could kill me.

"The real problem," he adds, lowering his voice, "is when somebody starts fiddling with the data."

The specter of online surveillance, therefore, worries him much less than that of data corruption. And this explains why so much time and so many resources in Estonia have been invested in its second brain, the online ID system. This platform, Ilves reminds me, creates ways to exchange data that are completely secure. "Your data is yours," as Sten Tamkivi assured me—but the information is really *yours* in Estonia only if the government is ensuring that nobody is tampering with it. The greatest service, then, of government-as-a-service is establishing data integrity.

The role of the sovereign in the digital twenty-first century, Ilves insists, is to guarantee our identity. He calls this a "Lockean contract" and describes it to me as the "new social contract for digital times."

Given that the original Lockean contract underpinning Anglo-American representative democracy operates on a series of mutual obligations between the government and its citizens, I wonder about the equivalent obligations for our digital age. "So what's our responsibility in this new social contract?" I ask. "If the government guarantees the integrity of the data, what do citizens need to guarantee in exchange?"

"It's a fully transparent system," Ilves explains. Just as it keeps government honest, he says, so it also keeps us honest. The government can, if it needs to, examine our data. So we all have to take responsibility for our own online behavior.

In the information-rich democracy being constructed in Estonia, he suggests, there can be no digital anonymity. Everything people do—from paying taxes online to ordering medicine to posting opinions to driving cars—is done under their own names. So, for example, Estonian newspapers are connecting the ID system to their bulletin boards, making it impossible to comment anonymously. This new social contract does away with the trolls who have made the internet such a barbaric place. And it makes people accountable for the spreading of fake news, racism and sexism, spiteful rumors, and the other antisocial behavior seemingly endemic in digital culture.

"Our goal is to make it impossible to do bad things without consequences. We want to teach people to be good on the internet, to use it responsibly," he says, sounding a little like a headmaster. I remember that Estonia is planning to establish a national-level test in schools to measure digital competency. I wonder if civic responsibility might be part of this exam.

In a way, of course, this is all very chilling—especially to people like Edward Snowden, who lionize individual privacy. It doesn't escape me that we are having this conversation in a replica of an eighteenth-century palace built by Peter the Great—the Russian conqueror of Estonia—in honor of his wife Catherine. Given Estonia's history of being ruled by supposedly "enlightened" despots such as Peter and Catherine the Great (not to mention Stalin), the country is all too familiar with the dangers of state control.

The Estonian ID operating system, with its guarantee of identity, is, in contrast, a mutually transparent system. What Ilves describes as a new social contract is based on the rights of both government and citizens to observe each other. The watching is done with full transparency, within a comprehensive legal framework that requires the authorities to alert people if they look at their data. It's an architecture of trust designed for what Andreas Weigend, the former chief scientist of Amazon, calls our "post-privacy" world.[17]

Moore's Outlaws

So what to make of Estonia? Does this country without borders offer a preview of our twenty-first-century fate?

Stuff might happen first in Estonia. But will it happen everywhere else next?

Perhaps. The Estonian model does come, however, with three important caveats. First, it's important to remember the country's ahistorical exceptionalism. Like other start-up nations such as Israel and, as we shall see, Singapore, Estonia has been able to reinvent itself because of its good fortune in being able to stand outside history. Just as Israel began in 1948 without any legacy institutions or traditions, so the post-1991 Estonian digital revolution occurred because a new generation of technologically literate policy makers and politicians—the Radio Free Europe journalist Ilves and his Princeton- and MIT-educated cadre of entrepreneurs and programmers—filled the vacuum created by the retreating colonial Soviet bureaucracy. Everything changed overnight. August 1991 marked the beginning of Year One in modern Estonian history.

Second, there is the distinctively unexceptional nature of the Estonian economy. Ilves, you'll remember, told me that Estonia was a "very poor" country when, in August 1991, the Soviets left. But even today it remains a relatively underdeveloped place, especially in comparison with advanced postindustrial economies like the United States, Germany, or Singapore. Estonia might top the trust league, but it doesn't lead the world in much else. A tech megabillionaire, a Zuckerberg or a Bezos, could probably buy little Estonia outright if he wanted. Its per capita GDP of around $17,600, for example, is ranked forty-second in the world (slightly above middle-rank economies like Russia and Turkey but a third of Singapore's $52,900), and the average monthly wage, after taxes, of its workforce of 675,000 is under a thousand euros. So reports of Estonia as the next Silicon Valley are, to be polite, slightly exaggerated. Ilves is under no delusions about this. While he believes the Estonian government-as-a-service experiment is "scalable," he acknowledges that it is more suited as a model for "developing countries" such as India rather than for advanced democracies.

And the third caveat is separating its appearance from its reality. All the policy makers and legislators with whom I spoke—from the green-jerseyed Andres Kütt to Linnar Viik, the country's Mr. Internet—have, in the best Silicon Valley fashion, drunk the Kool-Aid and thus loudly proclaim the triumph of their "country in a cloud." But the truth is less triumphant. This revolution remains a work in progress, and many ordinary Estonians remain indifferent to a lot of these digital abstractions.

Nonetheless, Estonia does matter. It matters because the government is prioritizing what Toomas Hendrik Ilves calls "data integrity"—the rather prim-sounding issue that will, surely, come to dominate conversations about twenty-first-century politics. It matters because Estonia is creating a model of digital government that is the antithesis of the one being created by its antagonist, Vladimir Putin's Russian Federation.

His vast eastern neighbor is never far from the Estonian president's mind. Between 1945 and 1946, Ilves reminds me, the Soviets "brutally occupied" the country and destroyed ten million Estonian books in an effort to completely obliterate the indigenous culture. And now Putin's Russia preoccupies Ilves. When, for example, I inquire about the purpose of the Estonian secret service, he explains that its "goal" is "to track down Russian spies." And while he acknowledges that the immediate threat of a Russian invasion of Estonia has receded since "the cyberattack that changed the world"[18]—Web War I, the notorious Russian cyber invasion of 2007, when Russian hackers shut down the Estonian internet—he certainly doesn't dismiss the threat of another invasion from the east, virtually or otherwise. Which is why, he explains, the Estonian government is digitizing all the country's books and written records and shipping them out of the country. And it's why, he adds, smiling grimly, "We are in NATO."

Rather than xenophobia or a desire for historical vengeance, however, Ilves's hostility to contemporary Russia is rooted in his profound distaste for a new category of government that Putin is pioneering. It's a twenty-first-century type of dictatorship, he explains, that has grown out of what Ilves dubs the "post-truth" philosophy of French postmodernists like Jacques Derrida and Jean Baudrillard. Orchestrated by Vladislav Surkov—Putin's personal advisor and, according to the Anglo-Russian writer Peter Pomerantsev, the "hidden author of Putinism"[19]—it's a style of government intent on transforming politics into a tightly produced reality television show of innuendo, gossip, and menacing unreality. Surkov's Putinism—not unlike, in many ways, the Trumpist spectacle produced by Stephen Bannon and Breitbart News in the United States—transforms politics into a real-time, always-on war of misinformation and disinformation. It is a definitively untrustworthy operation, run, to borrow some words from the Russian opposition leader Alexei Navalny, by a party of "crooks and thieves." And the internet, with its absence of curatorial authority and the anonymity of many of its users, has become the ideal medium for orchestrating this new type of war against truth.

"The real problem," Ilves warned me, "is when somebody starts fiddling with the data."

That's the essence of Putinism: *fiddling with the data.* Russia, explains Peter Pomerantsev, "reinvents reality, creating mass hallucinations that then translate into political action."[20] And this reinvention of reality relies on fiddling with the facts, particularly the digital facts, because—outside of the kind of meticulously secure system developed for Estonia's ID—they are so easy to manipulate, so easy to fake, so simple to turn upside down and transform into lies.

The British historian Simon Schama tweeted that "indifference about the distinction between truth and lies is the precondition of fascism. When truth perishes, so does freedom." Putin's whole operating system—with its reliance on opacity, fakery, and lies—is a metastasized kind of fascism. The Russian state is making a massive investment in the machinery of digital untruth as the main vehicle—a kind of digital Ministry of Truth—of its domestic and foreign policy. More than that, it has set up a four-story office complex in St. Petersburg as the headquarters of Russia's "troll army," the hundreds, maybe even thousands of bloggers paid by Putin to lie on the internet about everything from Hillary Clinton and Donald Trump to the war in Ukraine.[21] The Kremlin sponsors the online harassment of anyone who tries to expose these trolls and has, according to the *New York Times*, hired "elite hackers" and made cyberwarfare a "central tenet" of expanding its interests overseas.[22] Russia spends

$300 million annually on a "cyber army" of a thousand hackers, which, according to the *Financial Times*, is known as APT 28 and also goes under the name of Fancy Bears' Hack Team. It has been held responsible for interfering with the 2016 US presidential election.[23] Things have got so bad in this onslaught of fake news that in 2015 the European Union set up East Stratcom, its own eleven-person team to defend the Continent against fake news. Recent online lies have included the Swedish government's support of the Islamic State and EU's plans to regulate snowmen. Created by the EU to address, in its words, "Russia's ongoing disinformation campaigns," East Stratcom has discredited twenty-five hundred stories in the sixteen months since its establishment.[24]

A year after I met Ilves, I was invited to speak at the St. Petersburg Economic Forum, often dubbed the Russian Davos, on a panel about government policy toward data. The panel, which included two of Putin's most senior advisors on digital policy, was addressing the question: "Is big data a natural asset or a commodity?" As so often at these types of high-profile events, the platitudinous conversation was instantly forgettable. But had the panelists been honest, they would have invented a third category for big data in Russia. Rather than a natural asset or a commodity, big data—in Putin's Russia, at least—is becoming a weapon, something that enables the perpetual waging of war on your enemies.

Ilves may, therefore, be right in his criticism of Edward Snowden—who, of course, just happens to be exiled in Moscow under the protection of Putin's secret police. Worrying about individual privacy—with its now rather outdated Orwellian tropes about a Stalin-like Big Brother with a clear ideological agenda—might be very last-century. The new nightmare is the digital vertigo created by the fact-fiddlers in the Kremlin; the new nightmare is an endless misinformation war launched from the opaque depths of the Kremlin. And the only defense against this is what Ilves calls "data integrity"—a secure and transparent system that guarantees the reliability of information.

In their book about blockchain, the open-source public ledger technology that they call "the trust protocol," Don and Alex Tapscott observe that technology is "at the heart of just about everything—good and bad" in the world today. So Moore's Law, they warn, "doubles the power of fraudsters and thieves—not to mention spammers, identity thieves, phishers, spies, zombie farmers, hackers, cyberbullies, and datanappers." In Putin's Russia, digital outlaws have found a state-sponsored home, a place that will sponsor their fiddling with all our data—from the identity of our blood type to anything else that could harm us. In an age of accelerated evil, the Russian Federation—with its $300 million annual budget to support Fancy Bears' thousand elite hackers—has become the world's hub of digital deceit and disorder.

Estonia obviously can't fight the shadowy army employed by Putin and his cronies. But what this republic on the northwestern edge of Russia can do is offer an alternative model of a transparent, open, and fair kind of political system. One that is the antithesis of the monstrous purveyor of untruth that is emerging on its eastern border. One that prioritizes trust and is built upon the integrity of data. One, above all, that makes all of us accountable for our online behavior.

You Are Under Surveillance

That's why Estonia matters. It offers a glimpse into how to live well in cyberspace. And it's also become a model for other developing countries that are trying to create the kind of ID system Estonia has pioneered. In India, for example, the Estonian model has been extremely influential in the creation by the Modi government of a digital ID system known as Aadhaar, which is based on biometric and demographic data and is designed to give a digital identity to all 1.2 billion Indians.

One of the architects of Aadhaar is Viral Shah, a Bangalore-based technology entrepreneur with a PhD in computer science from UC Santa Barbara. Like so many of the technically brilliant young Estonians I met in Tallinn, Shah has leveraged his skills as a start-up tech entrepreneur to make the public sector more innovative and responsive. Together with Nandan Nilekani, the former CEO of Infosys Technologies, India's second-largest IT services company, Shah was commissioned by the Indian government in 2009 to "reboot India" by building a low-cost data ID system that would realize, in Nilekani and Shah's words, "a billion aspirations."[25]

"People in India are *dying* to be seen by the system." The youthful Shah explains the pre-Aadhaar India when we meet at his club in Bangalore. "If I'm a lower-caste person and get picked up by the police—I have no ID, no rights, no defense.

"So we need," he insists, "to get over the hump of identity."

Viral Shah sees many parallels between India and Estonia in getting over this *hump*. "Government is the most trusted brand in India," he explains, noting the similar political cultures in the two countries.

"What I love about India is that institutions work," Shah says, adding that there is a relatively high level of trust between politicians and the electorate. And the Aadhaar project, he explains, pointing to the similarities with Estonia, was used to "amplify" this trust.

Where India can learn from European countries like Estonia, Shah adds, is on the "burning issue" of privacy. There's no privacy law in India, he tells me, and there's a need to build what he calls "platforms of trust" between the people and government. Europe is "far ahead" on these regulatory issues, he acknowledges. Echoing Toomas Hendrik Ilves, Shah argues that there is a need for what he calls "social contract theory" to reinvent the relationship between citizens and the government in the digital age.

Other Indian technologists share Shah's concern over privacy. Sidharth Bhatia, a New Delhi–based journalist writing for the *Wire*, tells me that he's ambivalent about Nilekani and Shah's Aadhaar project. "I worry," he tells me over tea in a New Delhi mall, "because there are no checks on government." The New Delhi–based Arvind Gupta, the former head of the information tech unit at the ruling BJP party, also agrees. "It's very important to come out with a policy that addresses the protection of privacy," Gupta tells me, "if we want to be the biggest democracy in the world that is also digital."

Some Indian critics of Aadhaar actually use the Estonian ID system as a model for improving the Indian one. Sunil Abraham, the executive director of the Centre for

Internet and Society, a research organization based in Bangalore, tells me that India has much to learn from Estonia. Over fish curry in the garden of his office, he tells me that the Estonian system is better because it's based on internet technology rather than biometrics. Moreover, Abraham adds, Aadhaar needs what he calls a "decentralized authentication infrastructure"—something akin to the blockchain-like solution that Siim Sikkut and Linnar Viik have constructed in Estonia, which protects the privacy of its citizens.

Of course, India, with its 1.2 billion people, of whom only 35 percent are online, isn't Estonia. But the challenges that the two countries face as they try to reinvent themselves in the digital age—the challenges of trust and privacy and the reinvention of a social contract between the state and its citizens—aren't dissimilar. "Government not only *can* do good" in fixing the future, Viral Shah, the entrepreneur who rebooted India to realize a billion aspirations, tells me. "It has to."

As one drives out of New Delhi on the way to Agra, the home of the Taj Mahal, there is a large sign on the road. You Are Under Surveillance, it says. Estonia matters because it offers a country like India a model for establishing digital citizenship while protecting privacy rights.

Estonia: Where stuff happens first. But it's not, fortunately, the only place where stuff happens. There is, indeed, another place on the other side of the world that, like Estonia, is leading the world in digital innovation. It's an island— one not entirely unlike Thomas More's imaginary little place that resembled a grinning human skull.

CHAPTER FIVE

UTOPIA: A CASE STUDY
(BOOK TWO)

The Smart Island

The island of Utopia, we are told by Raphael Hythloday, the fictional guide to the no-place in More's little book, is crescent-shaped and resembles a new moon. It's possible that Thomas More got his geographical inspiration for Utopia from the archipelago of islands in the Strait of Malacca, the strategically valuable channel at the juncture of the Pacific and Indian oceans that had attracted the geopolitical attention of all the sixteenth-century European colonial powers. He might have even based it on a little place at the southern tip of the Malay Peninsula called Temasak. This island, also known as Singapura (meaning Lion City), was half the size of Utopia but with the same well-sheltered waters. It lay halfway between China and India and had been a commercial hub of Buddhist, Hindu, and Muslim traders for a millennium.

That part of the world would certainly have been familiar to somebody of Thomas More's international network, with his central role in the northern European Renaissance and his career as a lawyer and counselor to the king of England. In 1511, five years before More first published *Utopia*, the Portuguese had conquered the entire Strait of Malacca, including Singapura. A century later, the Portuguese burned the island community to the ground. This marshy, mosquito-infested island, with its 250 miles of coastline and its few hundred inhabitants of mostly pirates, was then occupied in 1819 by a British colonialist called Sir Thomas Stamford Raffles who, in the name of the East India Company, established what he called Singapore as one of the most strategically valuable possessions of the nineteenth-century British Empire.

Our interest in the Singapore story begins in 1965— the same fateful year that Intel cofounder Gordon Moore invented his eponymous law—with the establishment of the Republic of Singapore by a Cambridge University–educated lawyer called Lee Kuan Yew. Like pre-1991 Estonia, pre-1965 Singapore was an impoverished, ex-colonial backwater without natural resources, legacy institutions, international influence, or regional standing. Today, in contrast, postcolonial Singapore, like post-Soviet Estonia, has so successfully reinvented itself that it has become a byword for economic and technological innovation.

The similarities between these two countries on quite different sides of the globe—one about a hundred miles north of the equator, the other five hundred miles south of the Arctic Circle—are uncanny. Like Estonia, Singapore is today one of the most wired countries in the world, with its 152 percent mobile phone penetration, the highest on the planet, and its nationwide fiber network offering universal fiber-to-home internet access of up to ten gigs per second (that's a hundred times faster than the average broadband access in the United States). In the McKinsey Global Institute's 2016 *Digital Globalization* report, Singapore actually topped its Connectedness league as the world's leader in the inflow and outflow of goods, services, finance, people, and data.[1]

Like Estonia, Singapore also excels on the Edelman Trust Barometer, ranking among the world's top five countries for public trust in institutions in its 2015, 2016, and 2017 indexes. Like Estonia, Singapore has made programming and other digital skills one of the core tenets of its highly reputed educational system. And Singapore has also become a country undefined by its natural borders, with 25 percent of its current territory having been reclaimed from the surrounding sea since independence. "Stretching the land" is the euphemism Singapore planners use to describe the island's reinvention of its geography by flattening some of its hills and recycling that earth to fill in the surrounding sea.[2]

This tiny tropical island on the southern tip of the Strait of Malacca is charting a map of the networked future that has great relevance to the rest of the world. Just as Estonia is pioneering data integrity, so Singapore is the pioneer of what it calls a "smart nation." Launched by Lee Kuan Yew's oldest son and current Singapore prime minister, Lee Hsien Loong, in 2014, the state-directed "Smart Nation" initiative is designed to transform the island into what one minister calls a "living laboratory" for enabling data to improve citizens' lives. In 2016, the *Wall Street Journal* announced that Singapore "is taking the Smart City to a whole new level" and described the project as "the most extensive effort to collect data on daily living ever collected in any city."[3]

Lee Hsien Loong is designing a city like More's Utopia, where "there are no hiding places."[4] The utopian goal of the Smart Nation project is to record everything in Singapore so that everything about Singapore will be known by all Singaporeans. It's an attempt to actually realize the first law of geography: to make *everything* connected.

"Singapore is deploying an undetermined number of sensors and cameras across the island city-state that will allow the government to monitor everything from the cleanliness of public spaces to the density of crowds and the precise movement of every locally registered vehicle," explained Lee Hsien Loong at a political event in Singapore, using language that would understandably not only spook privacy advocates, who have nineteenth-century sensibilities about an individual's intrinsic right to be let alone, but also worry critics of Singapore's semi-authoritarian political culture.[5]

There is much for these critics—which include Amnesty International, opponents of the governing People's Action Party, and human rights groups—to worry about. In spite of all its remarkable economic achievements, the same People's Action Party has ruled Singapore without interruption since independence. While Singapore's constitution guarantees free speech, it also permits the government to control this speech with "such restrictions as it considers necessary or expedient" in order to protect "the privileges of Parliament" and to prevent "contempt of court" or "incitement to any offense."[6] And so, Singapore's constitution guarantees free speech as long as what you are saying isn't deemed offensive by the governing party. It's no wonder, then, that Reporters Without Borders, in its 2015 World Press Freedom Index, ranked Singapore 153rd out of 175 nations—sandwiched between Ethiopia and Swaziland and lower than Russia, Myanmar, or Zimbabwe.

Although the Singapore government has, according to the *Economist*, "begun expounding the virtues of thick skins," with several senior ministers, including Lee Hsien Loong, openly extolling the value of "naysayers" and "subversive challengers," the country's record on freedom of speech remains, at best, spotty.[7] In 2017, for example, the Singapore Supreme Court upheld a conviction against three activists who'd been protesting against a government-administered compulsory savings scheme. Most troublingly, the three protesters—one of whom had run as an independent candidate for Parliament in 2015—had been demonstrating in the country's Speakers' Corner, a space where Singaporeans are supposed to be able to express themselves without any restrictions.

A May 2016 criminal case against Amos Yee, a seventeen-year-old blogger charged with "wounding the religious feelings" of Muslims and Christians under the controversial Section 298 of the Singapore Penal Code, attracted the attention of Amnesty International. Two years earlier, Yee had been sentenced to fifty-five days in jail for mocking Lee Kuan Yew. Amnesty International has also taken up the cases of a couple of political activists who, in June 2016, were "subjected to hours of investigation" after posting something on Facebook on a day when campaigning was prohibited. "Amnesty International is deeply concerned by the continued sensitivity of the Singapore government to criticism and alternative views of Singaporeans," the human rights group wrote in June 2016.[8]

Rather than a digital Big Brother–style surveillance project to enforce the views of the People's Action Party, however, the goal of Smart Singapore—at least according to the government ministers, policy makers, and technologists who are building the program—is the creation of a civic intelligence platform designed for a post-privacy age. Indeed, if the E-stonia experiment is summarized by the prioritization of trust, then the point of the Singapore Smart Nation program—which also features an Estonian–style digital ID card system that stores an individual's personal data—can be captured by a focus on intelligence. The plan, we are told, is to reinvent Singapore so that digitized information—all those exabytes of data it produces every day—becomes the foundation of a supersmart new public space for civic innovation. The plan is to erect an electronic mirror that will enable the world's most connected country to become ever more self-aware.

Smart Nation is certainly an ambitious plan. As the five and a half million Singaporeans, 82 percent of whom use the internet and 64 percent of whom are active on social media, produce more and more data, so it will be funneled—via their smartphones, smart homes, smart vehicles, smart streets, even their smart schools—into the smartest of databases. Rather than a closed governmental monopoly of its citizens' information, however, the architects of the program insist that this Smart Nation database is being designed as an open platform, dubbed "Virtual Singapore," accessible to everyone, especially app-building entrepreneurs and civic technologists.

If, as the American ethicist Dov Seidman says, computers represent our second brain, then this ID system is designed to be Singapore's second brain. It is being designed as the database of national intentionality—a kind of digital commons of everybody's information—on which the networked society's new operating system runs. "Know thyself," the wise men of the ancient Greek city-states advised. Twenty-five hundred years later, a twenty-first-century city-state is transforming its collective knowledge into the steersman that will guide it into the future.

It's not surprising that superintelligence is the defining quality of Singapore's new digital operating system. The whole Singapore miracle is, in fact, built upon it. The city-state's founder, Singapore's King Utopus, Lee Kuan Yew, was—according to both Bill Clinton and Tony Blair—the twentieth century's smartest leader. Working without any natural advantages, Lee transformed this impoverished island on the southern tip of the Malay Peninsula into one of the world's most prosperous and orderly countries. Lee achieved what had previously been seen as unachievable—creating a one-party, paternalistic political system that combined a smart government committed to the public good and a comparatively incorruptible bureaucracy with an open, meritocratic culture of personal responsibility, hard work, and entrepreneurial innovation. What he did was use a top-down strategy of regulation, innovation, and education to create a communitarian, uniquely Southeast Asian formula constructed upon the most productive and innovative qualities of Singaporean citizens.

But there were also significant costs and compromises connected to this achievement. The "one shadow" hanging over Lee's legacy, noted Ishaan Tharoor, the *Washington Post*'s foreign affairs writer, in Lee's obituary, was his "draconian methods" of sometimes stifling democracy. "Under Lee, Singapore was governed as a virtual one-party state," Tharoor explains. "Freedom of speech, despite slow reforms, was strictly curtailed. Intense libel laws led to the bankrupting and marginalization of opposition politicians."[9]

There is, indeed, much that Singapore's semi-authoritarian founder, Lee Kuan Yew, shares with King Utopus, the unambiguously authoritarian founder of the aggressively legislated and highly regulated state of Utopia. Just as Lee brought many out of poverty, so Utopus, as More's Raphael Hythloday reminds us, "brought its rude, uncouth inhabitants to such a high level of culture and humanity that they now surpass almost every other people."[10] But there is one vital difference between Utopia and Singapore. In Utopia's profoundly undemocratic agrarian communist system, which, in some ways, is more reminiscent of North Korea than of Singapore, money was completely outlawed, and nobody was allowed to own anything. In Singapore, in contrast, although the state plays an important role in the country's economic life (80 percent of Singapore's population, for example, still live in government housing), the free market and the accumulation of personal wealth have always been actively encouraged, and there is a quasi-democratic electoral system, albeit controversially skewed toward the ruling People's Action Party.

Today, Singapore, the most connected place on earth, has under 2 percent unemployment and is ranked as the world's second-busiest port and the second-easiest place on the planet in which to do business. It now has a per capita income of $90,151, making it the third-wealthiest country in the world, after tax-free Luxembourg and natural gas–rich Qatar. By 2050, a Citigroup study predicts, it will be the richest country in the world, with an estimated per capita income of $137,710.[11] Singapore might not be quite Utopia. But in economic terms, at least, it's not far off either.

Lee Kuan Yew resigned as prime minister in 1990 and died in 2015, and his work is being continued by his son, Lee Hsien Loong, a Cambridge University–educated mathematician and computer scientist. But the Smart City initiative, rather than just a marketing strategy to maintain his father's legacy, is the next step in Singapore's journey from a sleepy colonial outpost to the world's most intelligent nation. The challenge of Lee Hsien Loong—who enjoys writing programming code in his spare time[12]—is not merely to digitize his father's successful formula and transform analog Singapore into an E-topia. It's also to eliminate that "shadow" hanging over his authoritarian father's legacy—that disdain for democracy—and ensure that the world's first smart nation is also a democratic one.

Geography Is Power

I am having afternoon tea with the writer Parag Khanna in Singapore's Goodwood Park Hotel. It's a colonial-style building, set in a lushly landscaped six-acre garden just off Scott's Road, the downtown street named after Captain William G. Scott, a nineteenth-century Englishman who had been Singapore's harbormaster and postmaster as well as the owner of some of the largest plantations on the island.

On an island in which everything is being perpetually reinvented, nothing in Singapore is ever quite what it seems. The Goodwood Park Hotel, for example, which was originally built by mid-nineteenth-century German settlers to resemble a Rhineland castle and was appropriated by the British and the Japanese as internment camps during the First and Second World Wars, has now been restored as the kind of idyllic British colonial estate once owned by Captain William G. Scott.

"It's a quiet place to chat," Khanna had emailed me when arranging where we should meet. The perfectly air-conditioned tearoom certainly is quiet, and cool too, compared with the noise and heat outside—the ninety-degree tropical weather, the crowded streets, the construction and traffic, all the endemic movement in this relentlessly industrious island city-state.

Khanna is the Singapore-based self-described "political geographer" with a "map obsession." His 2016 book, *Connectography*, tries to map a twenty-first-century "global Renaissance." Over Darjeeling tea, poured silently into fine china cups by an invisible waiter, the Indian-born, American-educated world traveler tells me why he had moved three years earlier, with his wife and two small children, from London to Singapore.

"You can't understand the future," he says, sipping his tea, "without embedding yourself in Asian life."

"How would this future look on a map?" I ask the author of *Connectography*. I think of the messages encoded in Hans Holbein's representation of Utopia. It occurs to me that, like the ever-changing Singapore, maps are never quite what they appear to be.

"It would be like Singapore," Khanna replies, waving his arm around the tearoom as if it were the whole island. "The geography of the future is a globally connected city."

Parag Khanna is right, of course. The vertiginous map of a hyperconnected Singapore—with its liquid inflows and outflows of networked goods, services, finance, people, and data—is, like it or not, ultimately all of our futures. But we know from Singapore's ability to "stretch" its land that geography and the making of maps are more than just the representation of physical reality.

After my tea with Khanna, I use my own rather crumpled paper tourist map of Singapore to navigate the city. On many buildings there are flags and placards celebrating the fifty-first anniversary of independence from Britain. HAPPY BIRTHDAY SINGAPORE—POWERING THE NATION these civic birthday cards all say. I take the gleaming metro downtown to the heart of Singapore—the Colonial District, the Marina Bay neighborhood down by the old quays beside the waterfront. It is here, amid its glittering collection of international banks, luxury hotels, and shopping centers, that Singapore's mastery of physical geography, its taming of nature, can be seen at its most grandiose. You can stroll in the Gardens by the Bay, the billion-dollar botanical gardens of the future, a conservatory featuring 217,000 plants from eight hundred species, including a grove of futuristic "supertrees" and a re-created mountain, complete with waterfall. And it's here you can shop, gamble, and sleep in Singapore's most iconic contemporary building, an $8 billion casino and hotel called Marina Bay Sands, a twenty-first-century wonder of the world that looks like something out of a science fiction movie: a cruise liner held up by three skyscrapers 636 feet in the sky.

My destination, however, is in the shadows of the main Marina Bay Sands building. Overlooking the waterfront, nestled like an overlooked shrub from the Gardens by the Bay, is a small, semicircular white building with ten fingerlike extensions and a bowl-shaped roof. Designed to represent the white petals of an open flower, this uniquely shaped edifice resembles one of the island's indigenous tropical plants—a giant white lotus.

Yet there is nothing indigenous about this building. Designed by Moshe Safdie, the same illustrious Israeli-Canadian-American architect who created the Marina Bay Sands (and also, more incongruously, Israel's Yad Vashem Holocaust museum), it is Singapore's ArtScience Museum, the city-state's leading venue for technology exhibits and festivals. I've come to the museum to see an exhibition entitled "Big Bang Data." It's the story of Lee Hsien Loong's Smart Nation project to master Singapore's virtual geography. Given that nothing in Singapore—especially its giant white lotus-shaped building—is ever quite what it seems, I'm here to try to distinguish the reality from the appearance of the Smart Singapore program.

The question I want to answer is simple. Is Singapore's second brain—this whole-of-the-nation endeavor to aggregate all its information in a centralized database—the solution or the problem for the future? Is it a "living laboratory" for a big data utopia, or is it a dystopia?

Digital Communitarianism

The "Big Bang Data" exhibition is ambivalent about the impact of data on society. On the one hand, it is full of warnings of mass surveillance, the overwhelming flood of online personal information, and the destruction of privacy. But the exhibition is also optimistic about the way in which data might enrich democracy and public participation. There are many different maps of Singapore, enough maps to satisfy even Parag Khanna, showing how the Smart Island is collecting and storing data in the cloud as well as providing more and more affordable access for its five and a half million citizens. In a section of the exhibition titled "Data for the Common Good," there are demonstrations of local Singapore apps designed for the public benefit. One, called Beeline, is a free transit app that seeks to improve the lives of commuters in Singapore by enabling the sharing of real-time data. A second app, developed by the Ministry of Health, involves ways that open data-sharing can facilitate communication between citizens and their doctors. A third, MyENV, created by the National Environment Agency of Singapore, offers real-time information about local air quality, weather, and dengue clusters.

I had been tipped off about the exhibit by Jacqueline Poh, the managing director of the Infocomm Development Authority of Singapore, the agency that is facilitating the development of the Smart Nation project in the education, health-care, taxation, transportation, and social services sectors. Over breakfast the previous day, Poh had advised me to go to the ArtScience Museum because, she promised, it would give me an insight into what she described as a "new era" in the inclusive way that technology is being delivered to the citizens of Singapore.

Poh, a petite woman in a smart red dress whose only visible concession to digital technology is the jet-black Apple watch she wears on her wrist, explains to me this new inclusive era. She expresses a vision of Singapore's digital future in the democratized language of a Silicon Valley visionary. She explains her agency's championing of open data platforms and hackathons that leverage Smart Singapore information for the common good. She boasts about civic hacking and the "cocreation" of apps between citizens and government. She tells me about Singapore's free, crowd-sourced MyResponder app developed by the Civil Defense Force, which enables the public to both report and respond to cases of cardiac arrest happening on the street. And she explains the importance of Data.gov.sg, the Singapore government's one-stop portal for the sharing of datasets from seventy public agencies.

But in contrast with the typically libertarian ideology of Silicon Valley, which fetishizes the free market and is deeply hostile to any governmental participation in the economy, Poh sees Virtual Singapore as an opportunity for citizens and the authorities to work together on projects that benefit both public and private interests. Whereas the seductive ideals of "sharing," "collaboration," and "community" are often appropriated by such successful Silicon Valley companies as Facebook to further their own intrinsically private goals, in Singapore these ideals fit naturally with the political culture created by Lee Kuan Yew.

Jacqueline Poh is an example of the sort of enlightened policy maker who is generating trust in the system. The daughter of a policeman—"We started out poor," she tells me—Poh studied politics, economics, and philosophy at both Oxford and Cambridge universities as a state-sponsored scholar before entering government service. She speaks about her "responsibility" to the *community*, citing what she describes as the "social justice" ideas of the American philosopher John Rawls. She is inspired by Rawls's thought experiment of a "veil of ignorance," which, she says, convinced her of a moral responsibility to look after the less fortunate members of society. Thus her support for Singapore's national broadband network, with what she calls its "comprehensive" fiber-to-home service for all citizens. And thus her pride in introducing the "Home Access" initiative for the lowest-income households in Singapore—a program that for just over $4 a month gives these households not only high-speed internet access but also a free smartphone.

Poh is far from the only civil servant in Singapore able to express her commitment to legislating the public good in the language of the Enlightenment. In the final room of the exhibition, there's another message, a kind of concluding poem for Smart Singapore, projected onto the wall:

WHEN THE SMART NATION SPEAKS, I'D LIKE TO THINK THAT IT IS ABLE TO TELL YOU STORIES FOR WHAT THAT FUTURE MIGHT LOOK LIKE
—AARON MANIAM/POET AND DIRECTOR OF INDUSTRY AT THE MINISTRY OF TRADE

The next day, I visit Aaron Maniam at the Treasury office in a downtown building adorned with HAPPY BIRTHDAY SINGAPORE—POWERING THE NATION banners. Like Poh, he's a smart young Singaporean from a typical middle-class family (his father was an air traffic controller) who won a government scholarship to study overseas. And like Poh, Maniam studied politics, philosophy, and economics at Oxford. He then spent a year at Yale and plans to go back to Oxford to write a doctoral thesis about the relationship between trust and technology.

"What," I ask the wiry, mustachioed young civil servant, "does data mean to you?"

It means, Maniam answers, the potential to bring a community closer together. And it means, he adds, the possibility to deepen the trust between the government and its citizens.

Like Poh, Maniam is interested in the philosophy of John Rawls. But he's also well versed in the work of communitarian thinkers such as Harvard University's Michael Sandel and the Scottish political philosopher Alisdair MacIntyre, who both prioritize the idea of the community over individual rights. Maniam tells me that he's particularly interested in the role that networked technology plays in building trust in the community and establishing a more interactive kind of democracy. Trust forms the basis of what he calls "social capital" that enables "human flourishing." He reminds me of Singapore's high standing in the Edelman Trust Barometer and cites what he describes as "deliberative platforms" like Salesforce's Chatter, which create a cooperative community of like-minded people.

He imagines a public version of Chatter designed as a community forum throughout Singapore. "I like to think of it as a commons," he says, using sixteenth-century English history to imagine the future of twenty-first-century Singapore. "It would be like the one Henry VIII closed in his destruction of the monasteries in sixteenth-century England."

So the Smart Nation, for Aaron Maniam, as it is for Jacqueline Poh, would be a digital commons for the cocreation of civic technologies that would form the basis of an interactive type of twenty-first-century technocracy. The sharing of data would be the foundation of a collaborative or communitarian-style democracy—a digital upgrade, in both form and function, of the uncorrupt quasi-democratic one-party system established by Lee Kuan Yew and maintained by his son Lee Hsien Loong. That's the story, they both imagine, of what the future will look like.

Trust, trust, trust. Though they are separated by several oceans and thousands of miles, it's striking how similar, in many ways, the thinking of senior public policy makers in Singapore like Maniam and Poh is to that of their contemporaries in Estonia like Taavi Kotka and Andres Kütt. What they all agree upon is the central role that trust will have in our post-privacy age. This view is shared at the very highest level of government too. Even Prime Minister Lee Hsien Loong believes that trust between citizens and government is the key to the success of his Smart Nation project. "People must be convinced that this is good for them," he says of the sixty-five thousand cameras that have been installed in public areas in Singapore since 2012—that "they will benefit from it, and it is not to intrude on their privacy."[13]

Lee Hsien Loong seems to be succeeding in this goal of building the trust of his citizens. In my time in Singapore, everybody with whom I spoke—from venture capitalists to start-up entrepreneurs to policy makers to technologists to the loquacious Uber drivers who drove me around the Smart Island—expressed an abiding, sometimes even almost eerie trust in the government. Maybe that's because, as Chinn Lim, the Singapore-based government affairs director at the Bay Area software company Autodesk told me, the government has choreographed a "fifty-one-year miracle" in the creation of a high quality of life, excellent education, and a plethora of jobs for its citizens. There's no hidden social contract in Singapore, Lim told me. To earn its responsibility, the government has always been accountable.

According to Jacqueline Poh, this trust emanates from the shared experience of the remarkable half-century journey that Singaporeans have collectively taken. "We all grew up together," the state policy maker says, describing a country in which public servants like herself and Aaron Maniam seem to trust Singapore's citizens as much these citizens trust them. Yes, this is the official line of the one-party regime in Singapore. And yes, there are some on the island—Amos Yee, for example, the teenage blogger jailed for mocking Lee Kuan

Yew—who would strongly disagree. But Singapore, in spite of its shadow of one-party authoritarianism, remains a trustworthy place. It ranks high on the Edelman scale. Citizens generally trust the government to look after their best interests.

So is the Singapore Smart Nation model, with its ambition to take data collection to a whole new level, the solution to or the problem for our digital future? It depends. As with More's Utopia, one is simultaneously impressed and chilled by this island city-state. The challenge is institutionalizing trust and creating what Estonian president Toomas Hendrik Ilves called a new Lockean social contract between government and citizens. And it's here that Singapore can learn from the Estonian model, with the guarantees of data integrity built into the system by technologists like Taavi Kotka and Andres Kütt.

Without this kind of institutional defense against untrustworthy government, the "living laboratory" that Singapore is creating could represent the foundations of a big data dystopia. After all, what happens in a hyperconnected society when trust between government and citizens is broken? What happens when an untrustworthy government knows everything about us? Most chilling of all, what happens if that government is not only untrustworthy but also untrusting—so that the very citizens it is supposed to be representing are all viewed as potential enemies?

Why 2020 Might Be 1984

"The big, big elephant in the room is the protection of privacy and ensuring security," acknowledges Vivian

Balakrishnan—Singapore's minister of foreign affairs and the minister-in-charge of the Smart Nation project— referring to the way that the Big Bang of data can undermine individual rights.[14] In the smart nation, Balakrishnan says, data integrity is essential. Without it, there can be no individual freedom in the digital age.

But when it comes to privacy, that elephant might not even be on the island of Singapore at all. Just as the antitheses of innovative Estonia are the state-funded militias of the Fancy Bears, with their $300 million annual budget, on its eastern border, so the dark side of the smart nation may lie to the north of the Strait of Malacca, in the People's Republic of China, a country with 730 million internet users and an untrusting and untrustworthy ruling class headed by President Xi Jinping. The problem is that Singapore's smart nation idea, with its aggregation of its citizens' personal data, might now be implemented by a totalitarian state without any respect for individual liberty. The problem is that China is seeking to create a smart national database where all intelligence will lie with the state rather than with the individual.

In contrast with Putin's Russia, Xi Jinping's China isn't an outlaw nation in today's hyperconnected new world. Whereas Russia's only real "innovation" is the fake news it anonymously exports to the rest of the world, China is one of the world's two leading innovation superpowers, with several indigenous winner-take-all internet companies that, in its own market at least, are crushing major Silicon Valley brands. Although part of the reason for this is a tightly regulated domestic market that discriminates against foreign companies, some of the success of these local internet companies—what the *Economist* calls "China's tech trailblazers"[15]—is related to the success of its own innovation ecosystem. China's dominant mobile-messaging service, WeChat, for example, has so elegantly combined email, free video calls, and instant group chats that equivalent American products such as Facebook Messenger or Apple Messages now appear slightly archaic.[16] China's e-commerce leviathan Alibaba and its successful microblogging website Sina Weibo have also added a selection of original features to make these products more popular in China than Twitter or Amazon. Uber, in fact, failed so dismally in China that it sold its local unit to its biggest indigenous rival, Didi Chuxing.

So the problem isn't about a lack of Chinese innovation. Instead, it lies in the way that this innovation is being manipulated by Chinese politicians and internet policy makers—the local equivalents of Singapore's Lee Hsien Loong, Jacqueline Poh, and Aaron Maniam. In contrast with the relatively enlightened and trusting digital strategies pursued by officials in Singapore, China's official internet policy is censorious, punitive, and both untrustworthy and untrusting. The building of the "Great Firewall of China," for example, has established a narrowly censored internet in which tens of thousands of foreign websites are blocked by the "Golden Shield," a censorship and surveillance project run by China's Ministry of Public Security. As in Russia, China has a not-so-secret two-million-person army composed mostly of government employees who flood social media with up to five hundred million pro-regime comments a year.[17] The Chinese government sometimes even shuts down hostile websites using the "Great Cannon," a tool specifically designed to launch distributed denial-of-service attacks. All too often, the Chinese government's heavy-handed attempts to solve its digital problems turn out to be even worse than the problems themselves. In response, for example, to the problem of youth internet addiction, a particularly serious affliction in China, the government created four hundred brutal rehabilitation boot camps that treated teenagers with electroshock therapy and other equally cruel punishments.[18]

In the preceding chapter I suggested that the Orwellian trope of a Stalin-like Big Brother is now outdated. But I'd overlooked China, where Orwell's twentieth-century warnings about a *Nineteen Eighty-Four*–style intrusive ideological dictatorship have acquired a twenty-first-century digital form. The idea began in 2010 with local Communist Party officials in Suining County in the Jiangsu Province north of the Chinese city of Shanghai. The local government began to award people points for good civic behavior and deduct points for everything from traffic tickets to "illegally petitioning higher authorities for help."[19] The points were then tallied, and people were placed into reputational tables to determine their civic reliability. If you rated highly, you would qualify for fast-track promotions at work or for access to public housing. If not, you were unlikely to be promoted, find a new home, or even qualify for official social security support.

In late 2016 the Chinese authorities published a plan to establish a much more ambitious social credit system that, in the Ministry of Truth–style language of the Communist Party, would build a culture of "sincerity" and a "harmonious socialist society."[20] Launched in three dozen local governments across China, as well as with eight private technology companies, and aided by increasingly sophisticated facial recognition technology,[21] this new initiative, called Internet Plus, is designed to collect the information of the 730 million Chinese internet users into regional databases that will determine their individual trustworthiness. The official goal is to unite all these different databases by 2020, creating a national social rating system that will rank individuals according to their online data.

Presumably Internet Plus, like the 2010 Suining County pilot project, will reward loyal citizens and punish those who are deemed, by China's intended second brain, to be politically troublesome or unreliable. The data-engineered caste system in the China of 2020 will, no doubt, be made up of two groups: the trustworthy and the untrustworthy. As Chinese officials put it, by 2020 this system will "allow the trustworthy to roam everywhere under heaven while making it hard for the discredited to take a single step."[22] The underclass, to remix Marx, will have nothing to lose but their bad ranking.

Trust, trust, trust. Just as the communitarian thinking behind Singapore's Smart Nation initiative is to increase the trust between citizens and their government, so the logic behind China's Internet Plus scheme is to tighten the grip of the Communist Party on society. In October 2016 President Xi Jinping called for innovation in "social governance" that would "heighten the capacity to forecast and prevent all manner of risks."[23] Rather than the creation of trust, the point of Internet Plus—which the *Economist* describes as a "digital totalitarian state"[24]—is to punish "untrustworthiness." To add to the Orwellian nature of this networked dystopia, individuals will be able to enhance their trust scores by informing on the untrustworthiness of others. Thus Internet Plus will, according to China's elite State Council, "forge a public opinion environment that trust-keeping is glorious" and, at the same time, "reward those who report acts of breach of trust."[25] The most trusted, then, in this surreal system, will be the most untrustworthy. No wonder one Hong Kong human rights activist told the *Wall Street Journal*, "It's just like 1984."

And yet in Singapore similar technology is creating an entirely different future. There's an important lesson here. It is a reminder that fixing the future is, for the most part, a political and social challenge rather than just a technological one. Technology doesn't solve technological problems; people do. Only humans can make their own digital histories. And this mostly happens, for better or worse, through governments. Some are mostly untrustworthy, as in China; others are more trustworthy, as in Singapore.

"When the smart nation speaks, I'd like to think that it is able to tell you stories for what that future might look like." These were the poetic words of Aaron Maniam in the last room of the "Big Bang Data" exhibition. But no nation, smart or otherwise, really "speaks" in a collective, Rousseau-like voice. Nations elect or appoint smart people—like Maniam or the Estonian president Toomas Hendrik Ilves—to speak on their behalf. Thus the most realistic stories about the future often come from legislators or regulators. And so it's to another prescient public official we must go next to get her story on what the future might look like. But she is located on the opposite side of the globe from Singapore—in Brussels, the European capital of regulation and legislation.

If Estonia and Singapore are, in their own different ways, countries in the cloud, then Brussels, with its gargantuan EU buildings filled with bureaucrats, is a place very firmly rooted on earth. It's time, then, to come down from the clouds. Time to talk to the regulators.

CHAPTER SIX

REGULATION

Reincarnating Teddy Roosevelt

I've flown from Northern California to the unloved, unlovely EU capital to meet the unlikeliest scourge of the Horsemen of the Apocalypse—a forty-nine-year-old Danish woman who grew up in a small town on the flatlands of West Jutland. Her name is Margrethe Vestager, and this former deputy Danish prime minister is now the European commissioner for competition—Silicon Valley's antagonist. More than any other single individual in the world today, Europe's antitrust chief is standing up to the business model and business practices of what, you'll remember, the *New York Times'* Farhad Manjoo calls the "Frightful Five"—the dominant winner-take-all tech companies whose collective $2.3 trillion valuation makes up an astonishing 14 percent of the entire $16.5 trillion GDP of the twenty-eight members of the European Union.

According to the *Financial Times* columnist Philip Stephens, Margrethe Vestager is the woman who is saving digital capitalism from the digital capitalists. The problem with free market capitalism, digital or otherwise, Stephens says, is its natural and perhaps even inevitable tendency toward winner-take-all monopolies. "Unconstrained, enterprise curdles into monopoly, innovation into rent-seeking," Stephens argues. "Today's swashbuckling 'disrupters' set up tomorrow's cozy cartels."[1] Thus his portrayal of Vestager as someone with the potential to be the twenty-first-century version of Teddy Roosevelt, the US president who deployed the 1890 Sherman Antitrust Act against industrial winner-take-all leviathans of the late nineteenth century such as Standard Oil and the American Tobacco Company.

But Roosevelt, Stephens reminds us, was no more of a socialist than Otto von Bismarck, the nineteenth-century German chancellor who laid down much of the groundwork for the modern welfare state. Rather than the redistribution of wealth, Stephens explains, the trust-busting Roosevelt recognized that "capitalism required legitimacy" and would thrive only if people had faith in a system that was fair to everyone. This was the same Teddy Roosevelt, you'll remember, who signed the Meat Inspection Act of 1906, one of the major pieces of legislation that, so to speak, cleaned up the meatpacking districts of New York and elsewhere. Roosevelt, like some of the other political fixers in this book, busted trusts in order to create *trust*. He made people believe once again in free market capitalism.

The whole point of trust-busting was to level the playing field for large and small entrepreneurs alike. Speaking in 1911 in support of a bill that would have set genuine limits on the size of corporations, the future American Supreme

Court justice Louis Brandeis observed, "There used to be a certain glamour about big things. Anything big, simply because it was big, seemed to be good and great. We are now coming to see that big things may be very bad and mean."[2] Most of all, Brandeis—who, you'll remember, also championed the right of the individual to be "let alone" by what he saw as the invasive technology of photography—believed that the excessive concentration of economic power in an America of Standard Oil and U.S. Steel was antithetical to democracy. "We must make our choice," Brandeis warned; "we may have democracy, or we may have wealth concentrated in the hands of a few, but we can't have both."[3]

Like Teddy Roosevelt, Louis Brandeis, and the many other early-twentieth-century Americans who reformed their industrial economy to make it fair once again for the small-business person, Margrethe Vestager is trying to rebuild trust in the economy. "It is too soon to pass Roosevelt's mantle to Ms. Vestager," Stephens admits, "but everyone who supports the liberal market economy that made possible the success of Apple, Google and the like should be applauding her courageous effort to reset the balance."[4]

Vestager's courage has been her unwavering determination to squarely stand up to what Farhad Manjoo calls the "new superclass of American corporate might."[5] By reminding companies like Google and Apple of their responsibilities in the real world, Philip Stephens suggests, Vestager—as it happens, the mother of three young girls—is socializing these often rather childish exponents of radical disruption. Apple's CEO, Tim Cook, for example, "often sounds as if he believes his company should be free to decide how much it pays in tax," Stephens notes.[6] And Vestager hasn't been shy about explaining to Cook his grown-up obligations. Thus her imposition of a 13-billion-euro fine on Apple in back taxes for what Stephens describes as the company's "labyrinthine tax arrangements" with the Irish government. Apple, Vestager reminded Cook, was paying just 0.005 percent tax on the tens of billions of dollars in revenue from its Irish-based European subsidiary to the Irish government. Given this brazen tax avoidance scheme, it's no wonder that Apple, the world's most valuable and wealthiest company, still holds $215 billion offshore, beyond the reach of any sovereign government.

I'd first met the redoubtable Vestager earlier that year in Munich at the annual Digital Life Design (DLD) Conference, Europe's premier tech event, organized by the German publishing conglomerate Hubert Burda Media. Speaking to an audience of mostly American technology entrepreneurs, pundits, and venture capitalists about what she described as an advertising-centric internet surveillance economy in which users' personal data has been turned into a commodity, the commissioner had argued that "consumers need a fair deal" to protect their digital privacy.

"Privacy," Vestager insisted in her Munich speech, sounding not unlike Edward Snowden or Louis Brandeis, "is a fundamental aspect of being." And freedom, particularly online freedom, she continued, should involve what she called "the right to be forgotten," making reference to an EU regulation that is part of an ambitious package to protect privacy.

Not surprisingly, given much of the audience's vested interest in a post-privacy economy, Vestager's presentation was greeted with only a smattering of polite applause. All the questions after the speech were hostile too. One well-known American tech pundit, a friend of Mark Zuckerberg and the author of an evangelical book about Facebook's power to unite the world, posed a particularly loaded question about the implications of Vestager's critique of Silicon Valley's core business model.

"Aren't you worried," he remarked in the morally pinched tone of a disappointed American internationalist, "about the fragmentation of the internet?"

Although the comment was dressed up in the holier-than-thou mode of a digital Wilsonianism, it was also a question about who would wield power and influence in the networked future. In Silicon Valley parlance, you see, "frag-mentation" translates into loss of market share. People talk of the "splinternet"[7]—the fragmenting, both for the better and for the worse, of the global digital economy into discrete regional markets—a concept deeply disturbing to private superpowers like Facebook and Google, whose shareholders demand endless growth and thus access to and dominance of a single world market.

According to the *New York Times'* Farhad Manjoo, this is a "fragmentation" that "is just a taste of a coming global freak-out over the power of the American tech industry." In the next few years, Manjoo forecasts, "we are bound to see increasing friction between the tiny group of tech companies that rule much of the industry and the governments that rule the lands those companies are trying to invade. What is happening in Europe is playing out in China, India and Brazil and across much of the rest of the globe, as well."[8]

Vestager's response to the DLD questioner was crystal clear. "No, I'm not worried," she said about the possibility that the European market might have regulations different from America's and that it could splinter off into a separate, even perhaps a rival economic ecosystem. "I'm not worried at all."

Congratulating her afterward for what I thought was a bravely uncompromising speech to a mostly hostile audience, I explained that I was writing a book about how to fix the future. "Ah, I might be able to help you with that," this twenty-first-century reincarnation of Teddy Roosevelt said with a little twinkle in her eye. "Visit me in Brussels, and we can talk more."

And so I do. But when I arrive in Brussels, I get the impression that war has just broken out. The gargantuan Berlaymont Building, the headquarters of the EU Commission's several thousand senior civil servants, appears to be on high alert. It seems as if all eighteen floors, forty-two elevators, and twelve escalators of this cruciform "Centre Administratif Europe" are under assault. There are armed troops, police vans, roadblocks, and various layers of EU security to negotiate before I am even able to enter this most immodest building at the Schuman Roundabout in the heart of the city's "European district." Had it not been for the terrorist attacks in Brussels the previous month, I might have assumed that all this exaggerated security was for Margrethe Vestager herself.

The commissioner, you see, was at war with Silicon Valley. Her face had been on the cover of the world's newspapers that week, with some cartoonists depicting Vestager as an ax-wielding Viking defending the European Continent from the American tech invader. It was the opening of a so-called third front in her regulatory war against Google that had triggered all these headlines. A few days before my visit, Vestager had signaled her intention to ramp up Europe's antitrust investigation against Google with a third formal investigation.[9] This dogged determination to hold US tech companies accountable under the EU law had even elicited the rather sour comment by the then American president Barack Obama that her efforts were "a form of protection-ism" designed to help European tech companies compete with Silicon Valley.[10]

Outside of Putin's Russia, where the indigenous search engine Yandex dominates, and Xi Jinping's China, where the Great Firewall blocks it, Google maintains an astonish-ingly tight stranglehold on the world's online information economy. Google's domination in the European market is even more exaggerated than in the United States, where Microsoft's Bing has around a 20 percent share of the mar-ket. In search, the Google query engine enjoys 95 percent of the market in Spain and Italy, 94 percent in France, and 93 percent in Germany.[11] But monopolies, in themselves, aren't illegal. Instead, what is illegal under antitrust law is for a company to leverage its dominance in one economic area to benefit another part of its business. In legal language, this is known as an "abuse of one's market position," and such a breach of antitrust law can have huge implications for two reasons. First, the fines can run into the billions of dollars (up to a tenth of a company's annual revenue, which in Google's case would be around 6.6 billion euros); and second, a ruling against a multinational leviathan like Google has the potential to reshape entire markets and industries.

"If a company is dominant, that's fine." Vestager thus summarized her antitrust strategy of guaranteeing a level playing field for smaller companies. "But if that dominance is abused, then we have an issue."[12]

And when it comes to Google, the world's second most valuable company after Apple, Vestager has many *issues*. Her multipronged investigations of Google are based on its alleged abuses of its dominant market position, particularly in online search and in its mobile operating system, Android. Google is accused of abusing its winner-take-all dominance in search by trying to control the entire online ecosystem. The first EU investigation is focused on accusations that Google abused its dominance in search by giving preferential treatment to its own price comparison service, known as Google Shopping—a classic breach of EU antitrust law.[13]

In June 2017, after a seven-year probe, Vestager's office fined Google a record-breaking 2.42 billion euros for this abuse of Google Shopping. "What Google has done is illegal under EU antitrust rules. It denied other companies the chance to compete on the merits and to innovate," Vestager said, explaining why she chose to fine Google so heavily. "And most importantly, it denied European consumers a genuine choice of services and the full benefits of innovation."[14] According to the British internet historian John Naughton, this "whopping" fine reflects a Silicon Valley "hegemony" that is now eroding. "The era when Europeans—and their governments—quailed before American corporate power may be ending," Naughton wrote shortly after the ruling.[15]

Vestager's second investigation examined the way Google was using its Android mobile-phone software—the operating system, you'll remember, used by 86 percent of the world's smartphone users—to strong-arm device manufacturers and telecommunications firms to install mobile Google as the default search engine on smartphones.[16] And the third is focused on the search engine company's use of its dominant AdWords service, which generates a substantial chunk of Google's annual almost $80 billion in advertising revenue. In this third case, the EU has, according to the *Wall Street Journal*, accused Google of "imposing restrictions on the way third-party websites display search ads from Google's rivals."

The Google advertising ecosystem, in other words, is attempting to be simultaneously an impartial platform for the distribution of advertisements and a corporate tool to crush its competition. Google is trying not only to become both church and state in the digital world but also to play God. Its goal is total control. It wants to own the full digital monty—control of *all* the online platforms, services, products, and stores in the networked economy.

God, however, might have met his match in a forty-nine-year-old Danish mother of three from a small town in West Jutland.

Silicon Valley's Dragon Slayer

If Margrethe Vestager is the reincarnated Teddy Roosevelt, then in Google she is confronted, in many ways, with the reinvention of the late nineteenth century's anti-innovation economy. It's a return to the industrial monopolies of John D. Rockefeller's Standard Oil, Cornelius Vanderbilt's New York Central Railroad, and J. P. Morgan's U.S. Steel—those "big things" that, Louis Brandeis suggested, "may be very bad and mean." That's why antitrust law is so important. Although it can be a rather dry, even arcane subject for nonlawyers, its importance for underwriting both innovation and fairness in our networked future can't be overstated.

Yes, as John Borthwick reminded me in our conversation at Betaworks, antitrust matters. Before I left for my trip to Brussels, I visited the Menlo Park law offices of Gary Reback, located in the very heart of Silicon Valley, just a few exits north on Route 101 from Google's Googleplex headquarters in Mountain View. Reback is about the closest thing America has to its own Margrethe Vestager.

"If there's one person who's going to help define antitrust law for the twenty-first century, it's Gary Reback," explained *Wired* magazine about Reback's outspoken support for a law that maintains the competitiveness of an open marketplace.

Indeed, twenty years ago, Reback played Vestager's role in a long-running legal drama that reinvented the future of the technology industry. In the late 1990s the Stanford-educated lawyer earned the moniker of Silicon Valley's "dragon slayer" when he convinced the US government to sue Microsoft because of its abuse of its dominant position in the desktop computing industry—with a 97 percent share of operating systems on all computer devices in 2000. Though not quite as abusive as John D. Rockefeller's Standard Oil or J. P. Morgan's U.S. Steel, Bill

Gates's Microsoft was a classic transgressor of antitrust law, intent on using its Windows operating system monopoly to crush perceived competition such as the Netscape web browser and Sun Microsystems Java technologies. Had Reback not taken on Microsoft, we'd probably still be living in a monopolistic Windows world. Even the internet itself might have become a platform for the promotion of Microsoft services and products.

Reback's aggressive advocacy was influential in the long-running legal case in which, you'll remember, John Borthwick also participated as the head of new products at America Online.[17] The ongoing drama of the three-year trial so distracted and weakened Microsoft that, even though the company ultimately wasn't broken up, it enabled the rise of the highly innovative Web 2.0 companies such as Google—the start-up founded in 1998 by a couple of Stanford computer science graduate students—which today has grown into the new global hegemon.

Reback—a fit, fast-talking attorney outfitted in the standard Silicon Valley uniform of khakis and an open-neck shirt—explains to me that the Microsoft suit was a "big deal." It was, he boasts, the largest antitrust case since the 1911 *Standard Oil Co. of New Jersey v. United States* confrontation, a legal saga that was so drawn out it filled twenty-one printed volumes of court records and eventually resulted in an eight-to-one Supreme Court decision in favor of splitting Standard Oil into thirty-four smaller companies.[18] Just as this early-twentieth-century trial represented Teddy Roosevelt's response to the challenge of new industrial technology, with its hitherto unprecedented rates of growth, so the

Microsoft case reflected what Reback calls the "network effects" that enabled Gates to transform his mid-1970s personal computer software start-up into the world's most powerful company by the mid-'90s.

"Back then," Reback confides, chuckling grimly, "Microsoft was so powerful that the government was quaking."

Reback prides himself on being what he calls a "Valley guy" in his uncompromising defense of technological innovation. What drove him to take on Microsoft, he confesses, was his distaste for the way it had tried to crush nimble start-ups like Netscape, the internet's first commercially successful browser, which was eventually acquired by AOL. "Start-ups," Reback insists, "have a right to present their innovative technology to consumers." And innovation, he says, is what drives both progress and economic growth.

So is history simply repeating itself? I ask, about the appearance today of new monopolists like Apple and Google.

Reback has a keen sense of history. In his very readable history of antitrust, *Free the Market! Why Only Government Can Keep the Marketplace Competitive*,[19] he explicitly compares what he calls the "excesses" of nineteenth-century "robber baron" industrial capitalism to the networked economics of the computer age. Yet Reback is too sophisticated a student of modern history to fall into the trap of seeing it endlessly repeat itself, like an annoying internet meme.

So yes, he acknowledges, Google is, in many ways, the new Microsoft, and the great challenge for antitrust lawyers like himself, as it was in the 1990s with the so-called Beast of Redmond, is to build a case against the Mountain View leviathan. But two things have changed over the last twenty years to make the Google case significantly different from that of Microsoft.

The first, he explains, is that Google's information technology is today "far more powerful" than Microsoft's software ever was. Compared with Microsoft's operating system or desktop publishing software, Reback tells me, Google Search, Google Maps, and the rest of Google's products and services have ten times more impact on individuals' lives. The company's technology is dramatically more "intrusive," he argues. Its business model, he explains, is "to build profiles of *everyone*."

Second, just as Google's technology is more powerful, Reback notes, so the Mountain View company is way savvier politically than Microsoft ever was. Perhaps, he suggests, it's because Google executives learned from Microsoft's arrogant mismanagement of its relationship with the government in the nineties. Or perhaps it's because Google is investing significant amounts of money—more than $15 million just in 2016, more than Dow Chemical, ExxonMobil, Lockheed Martin, or, indeed, any other internet company[20]—in its lobbying efforts in Washington, DC. This, Reback says, combined with the changes in the lobbying finance laws—particularly the 2009 *Citizens United v. Federal Election Commission* case that loosened rules governing campaign finance—has made it impossible for Google's critics to get the attention of government. In today's pro-corporate political environment, Reback insists, it's simply not possible to bring a federal antitrust investigation against the Mountain View giant.

Reback told me this in 2016, before the election of Trump and the establishment of an even more pro-business, antiregulation regime in Washington. Some people have predicted that Google "may face antitrust scrutiny under Trump"[21] because of Obama's unhealthily close relations with the Mountain View company, particularly his friendship with its executive chairman, Eric Schmidt. But given Peter Thiel's influential role in the Trump administration, especially as an advisor of the new regime's tech policy, this seems extremely unlikely. Thiel, the libertarian multibillionaire investor who made one of his most successful counterintuitive bets when, alone in Silicon Valley, he backed Donald Trump for the presidency, is an outspoken supporter of monopolies. In his bestselling libertarian manifesto *Zero to One*, Thiel even seems to suggest that it's the government's role to let monopolies alone because of their supposed efficiencies and wealth-creating benefits.

Reback, of course, would strongly disagree with Thiel about the collective benefits of monopolies. That's why he's so disconcerted about the American government's unwillingness to take on winner-take-all economics. "We can't make a difference anymore in the United States," he says, shaking his head ruefully. "We just don't have enough money."

And that's why, he believes, Margrethe Vestager's efforts in the EU are so critical. Antitrust law, with its implications about what the American lawyer calls a "common market," is much more relevant, he believes, in Europe than in the United States. And so Vestager's filings of both the search and the Android antitrust investigations, Reback says, are important steps in bringing Google to account.

"If change is going to come in the United States," he concludes regretfully, speaking of an America that combines a remarkably innovative economy with an increasingly dysfunctional political system, "it's going to have to begin in Europe."

It's also likely that a rethinking of antitrust law will originate in Europe, where regulators like Margrethe Vestager are under less pressure from the increasingly powerful tech lobby. As the *Economist* argues, "antitrust authorities need to move from the industrial era into the twenty-first century" in order to reboot antitrust for the information age.[22] Rather than size, the *Economist* wisely suggests, regulators should now take into account "the extent of firms' data assets when assessing the impact of deals." In this context, for example, Facebook's $19 billion acquisition of WhatsApp in 2014 would, according to the *Economist*, raise regulatory "red flags" because of the massive amounts of data involved in the deal.

It is also likely that a tech-savvy regulator like Vestager will begin to think differently about Amazon in the context of twenty-first-century antitrust law. As Amazon—with its rapidly growing e-commerce and Web Services businesses—increasingly becomes a utility providing the tools that enable digital business, so it will come under the scrutiny of antitrust regulators. "If Amazon does become a utility for commerce," the *Economist* predicts, "the calls will grow for it to be regulated as one." And as Amazon's sales and profits grow, regulators will, no doubt, become increasingly concerned with its unprecedented economic power. If Amazon does indeed make as much money as its bullish investors hope, the *Economist* speculates, its earnings could be worth "the equivalent of 25% of the combined profits of listed Western retail and media firms."[23] And if Amazon gets anywhere close to this dominance of the market, then it will also inevitably attract the scrutiny of regulators on standard antitrust grounds, particularly in Europe.

Gary Reback certainly isn't alone in seeing Vestager's work at the EU as being critical to keeping the digital marketplace competitive. While I was in Singapore, I met with Toh Han Li, the chief executive of the Competition Commission of Singapore. Google has 80 percent of the search market in Singapore, and as Toh Han Li—who favors what he calls "data portability" underwritten by the public sector—acknowledged, "data monopoly by a private company can be a dangerous thing." There had been a private action in the Singapore courts examining Google Maps in terms of antitrust, Toh Han Li told me. But a small agency like his, with its approximately thirty lawyers and thirty economists, doesn't have the resources, he admitted, to take on Google's might.

And so Toh Han Li, like Gary Reback, is waiting for Margrethe Vestager to take the lead. Where she goes, they will follow. What she accomplishes, therefore, will help fix the future not only in Europe, but also in the United States and Singapore.

In a memorable exchange in their October 2015 Democratic Party presidential primary debate, Bernie Sanders told Hillary Clinton that America had much to learn from Scandinavian societies. "We should look to countries like Denmark, like Sweden and Norway, and learn what they have accomplished for their working people," Sanders argued.

"We are not Denmark," Clinton lectured Sanders with the kind of tone-deaf arrogance that may have cost her the

2016 election. "I love Denmark. We are the United States of America, and it's our job to rein in the excesses of capitalism so it doesn't run amok."[24]

But Clinton was wrong. In a strange way, Denmark is leading the charge against the worst excesses of free market capitalism, as manifested in the immoral and arguably illegal behavior of the world's two most valuable multinational companies: Apple's brazen skirting of European tax laws and Google's attempts to corner the entire digital media economy.

A Danish Room of Her Own

Up in Margrethe Vestager's office on the fifteenth floor of the Berlaymont Building, away from the armed troops, police vans, and antiterrorist roadblocks out on the Brussels street, it's anything but a war zone. The apartment-size room is filled with abstract art, colorful rugs, a couple of comfortable sofas, and a sideboard full of photographs of her family, mostly of her schoolteacher husband and her three young daughters. It is as if she's carved out a little bit of Denmark inside the Centre Administratif Europe. This is the commissioner's refuge from unlovely Brussels. It is, as the Brussels *Financial Times* correspondent noted, a room that "feels *hygge*"—the modish Danish word meant to describe a feeling of cozy togetherness.[25]

But for all its Danish coziness, Vestager's office may be the most important place on the planet for determining the future of the global technology industry. As I sink into one of the sofas, I remember that this was the same room—and perhaps even the same seat—from where, a couple of months earlier, Tim Cook had appealed to Vestager's "fairness" in dealing with the dispute over Apple's back taxes owed to the Irish government. According to one of his associates, it had been "the worst meeting" they had ever experienced in Brussels. After the acrimonious encounter, Vestager fined Apple 13 billion euros, a punishment ten times larger than the EU's previous record for this kind of corporate crime. The normally unflappable Cook had angrily responded that the commissioner's decision was "political crap."[26]

Margrethe Vestager might be the trust-busting reincarnation of Teddy Roosevelt, but she certainly doesn't look like the Bull Moose. Wearing a sunny yellow dress, she is as warm and friendly as her office. Over tea, served with much less formality than in Singapore's Goodwood Park Hotel, the unelected commissioner explains to me her philosophy of government. She works on behalf of 507 million European citizens, she says, stressing her sense of responsibility toward every EU citizen. Reminding me that her mother still has a small store in the western part of Denmark, she says that, as a politician, she aims to give ordinary Europeans like her mother a "fair chance to participate and form their own lives.

"The good society," she says, raising her teacup in a mock toast, "allows each citizen to pursue his or her own dream."

She believes that there's a need for balance between the market and the government. "Without regulation, without law enforcement, the only thing that exists is the law of the jungle," she says, repeating the warnings of Karl Polanyi, the author of *The Great Transformation*, about the inegalitarian consequences of that great utopian seduction—a perfectly free market. If the markets are left completely unfettered, she explains, you get "winner-take-all" companies like Apple, Amazon, and Google. And a totally unregulated market, she adds, echoing the Silicon Valley antitrust lawyer Gary Reback, offers no protection for the start-up entrepreneur.

Yes, the free market has an important role in the future digital economy, she acknowledges. And yes, her regulatory work is focused on enabling internet innovation, particularly of the start-up entrepreneur, in the marketplace. "Google was a start-up, once," she reminds me. But the market, she insists, can't do the full job. It requires the oversight of regulators like herself.

She's critical of the online economy in which consumers are exchanging their personal data for free services and products. One solution to this surveillance economy is to reintroduce a more traditional form of monetary exchange and get people to once again pay for newspapers and other curated forms of online content. Another solution, she says, is for entrepreneurs to work on digital products that are specifically built around the guarantee of privacy—an architecture she describes as "privacy by design." The problem, the commissioner insists, is that there's no such thing as a free lunch—on or off the internet. So all these "free" online services like YouTube or WhatsApp are actually being paid for by consumers with their data as de facto currency. This isn't just an internet problem. She tells me that she even gave up her loyalty card at a Belgian supermarket because she doesn't want to give up her personal data. They "know everything," Vestager says about such stores. "And all you get in return," she sniffs dismissively, "is a discount on washing detergent!"

Like More's Raphael Hythloday, his fictional guide to the island of Utopia, Vestager loves to travel. "I'm the travel agent of my family," she tells me proudly. "Transportation is fascinating," she says of a sector that she hopes could become a laboratory for a civic experiment in the sharing of digital information. She imagines what she calls a "data commons" for Brussels in which people share their information about traffic and other travel-related issues. The problem with the travel app economy, she says, repeating Tim Berners-Lee's critique of the contemporary web, is that the data gets locked up in private silos such as Uber, Airbnb, or Lyft. So the information experiment she imagines for Brussels would be built on public infrastructure and would require the open sharing of all data by everyone.

Her commitment to More's Law, that moral instruction to serve one's community, is as unabashed as that of anyone featured in this book. In terms of increasing the amount of trust in the system, Vestager believes that we need politicians both to be more human and also to be more inspiring leadership models. Like that of policy makers in Estonia and Singapore, her goal is the rebuilding of trust between rulers and the ruled. The difference, though, is that while the Estonians and Singaporeans are creating digital technology to enable this reinvention, Vestager is relying on the traditional analog strategy of rebuilding trust by presenting herself as a role model of a trustworthy politician.

Her responsibility in this role is not to "service incumbents," Vestager insists, so sometimes it's necessary to aggressively use one's powers and not withdraw from conflict. Some people imagined, Vestager confides, that these Silicon Valley tech companies were too big to be taken on. But all she is doing, she explains, is siding with her fellow 507 million Europeans against multibillion-dollar multinationals like Apple, which cheats on taxes, and Google, which illegally crushes its smaller competitors. Tim Cook might call this "political crap," but for Vestager it's the essence of her calling as a public servant responsible for pursuing the interests of her community.

Not everyone, of course, favors this kind of socially interventionist government. Tim Cook certainly doesn't. Nor did Lee Kuan Yew, the founder of Singapore, who viewed it as foreign to his ideals. "Westerners have abandoned an ethical basis for society, believing that all problems are solvable by a good government," he said about Vestager's style of European social democracy. "In the West, especially after World War II, the government came to be seen as so successful that it could fulfill all the obligations that in less modern societies are fulfilled by the family . . . In the East, we start with self-reliance. In the West today, it is the opposite. The government says give me a popular mandate and I will solve all society's problems."[27]

Bernie Sanders, you'll remember, was more complimentary about this style of socially responsible government. And in contrast with Singapore, which became independent only in 1965 and thus had fewer legacy institutions or traditions, Europe—and specifically Scandinavia—developed a social welfare system in the late nineteenth century. It was a direct response to the social problems—the displacement of

rural communities, the urban slums, the unsafe factory working conditions, the working-class unrest—that represented
the early free market history of industrialization. And the
so-called Scandinavian model, as it evolved, featured relatively high levels of public expenditures, high-quality public
services, and a significant level of direct state participation,
including interventionist politicians like Margrethe Vestager
unashamedly working for the common good.

Before we tar all American tech entrepreneurs with
the same utopian laissez-faire free market brush, it's also
important to remember that there are some who agree with
Bernie Sanders about the role of government in stimulating
innovation. For example, Steve Case, the founder of AOL
and the most celebrated internet entrepreneur of the 1990s,
now believes in what he calls a "Third Wave" of innovation,
in which government needs to play a much more central
role in the digital economy.

"My view is simple," Case predicts. "Government is
going to be central in the Third Wave."[28] America needs
to learn to be more consensual, Case acknowledged to me
when we met in San Francisco. "We really do need to learn
from the German and Scandinavian examples."

There is some evidence of the emergence of Case's
Third Wave of innovation in America. Much of this is
being developed around the combinatorial idea of bringing
together innovators and regulators to enable better public
services. The chief technology officer of Los Angeles, Peter
Marx, for example, is pioneering the notion of the government as the platform for innovation in the transportation industry. The former CTO of Universal Studios, who

has spent his whole life in the private sector, insists that he isn't "a typical bureaucrat." As the CTO of California's largest city, Marx is pioneering the idea of what he calls "the city as a digital platform," in which Los Angeles publishes all its anonymized travel "data sets" on a GoLA app—every arrest, the location of every streetlight, and every parking ticket—for the benefit of the community. What he calls this "robust repository" for open data is, he says, "the exact opposite world to Facebook." He is building a public stack of technology—which he calls a "trustworthy layer"—above such private companies as Uber or Lyft. It's like the Smart City applications that are being developed by Jacqueline Poh's team in Singapore.

Marx tells me that the GoLA app provides the "infra-structure for good citizenship." What he is doing is taking what he calls "city infrastructure" and making it available for everyone. It is, he claims, the first urban mobility marketplace designed to help people get around Los Angeles quickly and more cheaply. With GoLA, Marx claims to be "creating the commons." He calls it a "classic California utopia that is actually realizable." It is, indeed, Margrethe Vestager's dream of a digital public layer of transportation data designed for the benefit of the community. How ironic, then, that a digital commons is being pioneered in Los Angeles by a lifelong technology executive from the private sector.

This ideal of the digital commons is emerging as one of the most politically sensitive issues of the early twenty-first century. The inviolability of digital public space has become captured in the controversial idea of "network neutrality"— the notion that governments or private companies should be required, under law, to treat all data on the internet equally. Lee Kuan Yew might have claimed that "in the East, we start with self-reliance," but in contemporary India, at least, the law requires the government to protect the independence of the digital commons. In India, network neutrality legislation even resulted in the first national uprising against one of Silicon Valley's private superpowers. In February 2016 India outlawed Facebook's Free Basics, a supposedly "free" internet service that promoted certain apps for the benefit of Facebook rather than that of the community, thereby besmirching the ideal that all Indians have the same access to the same internet.

On my trip to India, everyone I met—from venture capitalists to start-up entrepreneurs to technologists—supported the principle of network neutrality as a model for protecting the public good. But some were uncomfortable about the government taking unilateral responsibility for this protection. One former Bangalore-based engineer, Sharad Sharma, has created a nonprofit network called iSPIRIT, designed to build a coalition of technologists and businesspeople who can create products that Sharma, quoting the American psychologist Martin Seligman, says will promote "meaningful happiness." Sharma tells me that iSPIRIT, which is supported by both the Mozilla and the Kauffman foundations, has been funded by ninety-five donors, all by invitation. It specifically doesn't take money from venture capitalists, the government, or corporations. The point of iSPIRIT is to "build public digital goods without public money." Sharma's vision is to turn government into what he calls a neutral "platform" for the creation of technology for the public good. "All innovation is combinatorial," Sharma tells me. It can't be created from above by the government.

Sitting in Vestager's office, I remind her of her comment at the Munich DLD Conference about the splinternet—the fragmentation of the internet into a series of different markets working on different economic rules and cultural assumptions.

We already have a European zone, she reminds me, shrugging. I take her shrug to mean that there are no universal solutions; there is no single operating system that can solve all the world's digital problems.

"But there *is* a difference between EU and US societies," Vestager insists. "The tradition of the European welfare state *is* relevant. We Europeans also think differently from Americans about the idea of equality. Our behavior reflects our culture."

It's what we might dub the revenge of geography. The early ideal of the internet, enshrined so nobly by Tim Berners-Lee in his World Wide Web, was for a global electronic network that would unite humanity beyond all the differences of analog geography. However, in the world that Margrethe Vestager describes, more than a quarter of a century after Berners-Lee donated the web to the world, it's clear that technology is actually maintaining and sometimes even deepening traditional cultural divisions.

The internet began in America as an unholy marriage of self-interested free market capitalism and idealistic Wilsonian internationalism. Today, as we've seen in Estonia, Singapore, the EU, India, and even Los Angeles, it's being reinvented and refined in places that have quite different traditions. Maybe this book should really be called *How to Fix the Futures.*

The Great Digital Divergence

Margrethe Vestager is, of course, right. For better or worse, her vision of the future is a particularly European one, reflecting faith in "good government" as the only agency that can, in Lee Kuan Yew's words, "solve all society's problems." And it's a philosophy now being deployed across Europe in confronting the most troublesome aspects of the digital revolution—among them illegal monopolies, fake news, nonpayment of taxes, the unregulated flow of data back and forth across the Atlantic, and, above all, the threat to an individual's right to digital privacy.

It's the great digital divergence—Farhad Manjoo's "fragmentation" and John Naughton's erosion of hegemony—between Europe and America. Or at least between American Big Tech and European governments. According to Martin Schulz, the former president of the EU Parliament, the Frightful Five are pursuing a kind of radically disruptive technological agenda that is acting as a "wrecking ball" to culture and society.

"The aim is not just to play with the way society is organized, but instead to demolish the existing order and build something new in its place," Schulz says of these supposedly destructive tactics of Facebook, Amazon, Google, and Apple.[29]

And so, like the Danish room Margrethe Vestager has carved out in Brussels, the EU's digital future is being designed by politicians like Vestager and Martin Schulz as a cozy refuge from the Silicon Valley wrecking ball.

Vestager's so-called war against Google has now spread to many fronts, with the company's advertising-centric business model increasingly under regulative threat. In May 2016 French prosecutors raided Google's Paris office as part of their investigation into claims that 1.6 billion euros of back taxes are owed to the French government.[30] Google's online advertising business is facing much stricter privacy rules from Brussels on how it tracks people online, including proposed legislation that would require users to "opt in" if they are to be exposed to online advertising from Google.[31] The EU is even considering a tax on online snippets, the so-called Google tax, which would update online copyright law so that Google would have to pay newspapers and publishers for running short extracts of their content on its search engine and in its Google News service.

Given the threat of fake news from Putin's trolls in the critically important 2017 French and German elections, European regulators have moved aggressively to make the major social media platforms, such as Facebook and Twitter, accountable for the lies published in what two French journalists have dubbed *la fachosphère*. In January 2017 the former Estonian prime minister, Andrus Ansip, who is now the European commissioner for the Digital Single Market, warned Facebook and other social media sites that the publication of recent fake news, such as the viral lie that the pope had endorsed Donald Trump for president, represented a "turning point" in the credibility of online media. Ansip urged these social media companies to take more responsibility for their actions. Self-regulation was essential, Ansip said, if these platforms wanted to remain trusted.[32] He is open to other solutions too. The only fix that the EU fortunately isn't considering is active censorship of social media. As Ansip ruefully noted, perhaps borrowing from the experience of his country under Soviet occupation, "fake news is bad, but the Ministry of Truth is even worse."[33]

As the *Economist* drily reminds us, "it's no longer 2005"— and in a world in which violent fanatics of every stripe are actively recruiting on the internet, "legislators must strike a balance between security and liberty."[34] Measures sterner than self-regulation are certainly needed. "Facebook will be driven to remove content only if it could hurt its profit margin," notes one digital rights activist in Brussels.[35] And this is supported by a May 2017 *Guardian* report that quotes an internal Facebook document saying that moderators should block or hide Holocaust-denial material only if "we face the risk of getting blocked in a country" or face "a legal risk."[36] The *Guardian* found that Facebook actually hides or removes Holocaust-denial material only in the four countries where it fears it could be sued for the publication of this content: France, Germany, Israel, and Austria. So a private superpower like Facebook seems to respond only to realpolitik. The only way, then, to make Facebook acknowledge that it's a media company is to treat it like a media company and sue it for publishing illegal content. Appealing to Facebook's conscience and expecting it to do the right thing is, in contrast, mostly an exercise in wishful thinking.

Thus the German government's introduction of legislation in October 2017 that fines both social media sites and search engines up to $57 million if they don't delete illegal, racist, or slanderous material within twenty-four hours of its being tagged. The purpose of this new law, according to Heiko Maas, the German justice minister, is to provide the same punishments online as exist for illegal off-line speech. "With this law," Maas explained a few months before the law was due to go into effect, "we put an end to the verbal law of the jungle on the internet and protect the freedom of expression for all."[37]

Other European countries are also active in trying to curb fake news. The British Parliament has set up a House of Commons Select Committee to "grill Facebook executives" over their role in the fake news ecosystem.[38] Even the Czech Republic, which is also holding general elections in late 2017 and has been subjected to a relentless information war no doubt orchestrated by the anonymous Fancy Bears in Moscow, is planning to open a specialist fake news unit ahead of the vote to tackle falsehoods purposely spread on the internet.[39]

The growing global furor, with its threats of financial punishment over social media's publication of fake news, bestiality, child sexual abuse, and beheadings is, fortunately, finally having an impact. In February 2017, Facebook began working with Correctiv, a Berlin based nonprofit media start-up that is developing fact-checking software for exposing fake news.[40] Then, responding to an April 2017 Cleveland murder video that stayed up on Facebook for hours before being removed and a Facebook video of a Thai man killing his eleven-month-old daughter, the social media behemoth finally took measures to actively police its content. In May

2017, in response also to the announcement of the punitive new German law around online hate speech, Mark Zuckerberg announced that Facebook would hire three thousand more editors—supplementing the forty-five hundred gatekeepers already on this team—to review the user-generated content.[41] Now all Facebook needs to do is pay these editors a fair wage for such an important and emotionally draining job. "We were underpaid and undervalued," one Facebook editor complained about a job in which he was paid $15 an hour to, in his words, "turn on your computer and watch someone have their head cut off. Every day, every minute, that's what you see. Heads being cut off."

On the privacy front, officials in many European countries, including France, Spain, Holland, Belgium, and Germany, are conducting long-running investigations into Facebook's frustratingly opaque relationship with its users' data.[42] The EU is also clamping down on US messaging apps like the Facebook-owned WhatsApp and Jaan Tallinn's Skype, now owned by Microsoft. In August 2016 the EU announced its intention to extend the privacy rules covering traditional telecom services to apps like WhatsApp and Skype.[43] The following month, the German government ordered Facebook to stop collecting data on Germany's roughly twenty-five million WhatsApp users.[44]

In 2010, Max Schrems, an Austrian graduate student, while researching a thesis on European privacy law, asked Facebook to send him all the data it had attached to his account. It sent him a 1,200-page pdf containing records of all the IP addresses he had used to log in, in addition to information on which other Facebook accounts had logged in with that same machine. The report also included records of all the messages he had exchanged, all the "pokes" he had ever received, and even items he'd thought he'd deleted and personal information he believed he never provided.[45]

In 2013, Schrems filed a complaint against Facebook Ireland, the European headquarters of the company. And in 2015, he won a landmark privacy case at the European Court of Justice that invalidated a fifteen-year pact between the EU and the United States called "safe harbor," which enabled the free flow of transatlantic data. The *Financial Times* described the Schrems victory—which was accompanied by a herogram from Edward Snowden—as a "watershed moment in the transatlantic digital relationship."[46] What is authored in Europe, at least on the internet, now must stay in Europe. The separation between the EU and the United States, at least when it comes to the free flow of data across the Atlantic, is now a formal divorce.

None of these new laws or regulations compare, however, in their scope or ambition with a new privacy bill known as the General Data Protection Regulation (GDPR). This is the most important of all the EU attempts to duck the Silicon Valley wrecking ball. Adopted in April 2016 by the European Parliament, the European Council, and the European Commission after four years of negotiation, the GDPR will come into force in 2018 and has been designed as a single set of rules for member countries to ensure that privacy will become "the norm" in European networked society. Offering to guarantee the "fundamental right" to personal data protection for all EU residents, the GDPR is an attempt to turn today's data equation on its head by enabling individuals to own what technologists call their "social graph." Instead of big data companies owning us, we will not only own our data; we will be able to delete it or take it with us wherever we go on the internet. Google will no longer be allowed, in Gary Reback's words, to unilaterally "build profiles of everyone." Some are calling for a similar "Social Graph Portability Act" to be created by the US Congress.[47] They would be wise to use the intended GDPR act in the EU as a model for this congressional legislation.

The GDPR's "privacy as the norm" legislation offers an entirely new ecosystem for privacy and data. While Singapore's Smart Nation initiative remains a bit fuzzy about who exactly owns all the personal data on the internet, the GDPR is crystal clear. It's ours. Ours alone. The internet is reimagined as a place where individual privacy is not only the norm but also the highest priority. While the current absence of clear laws puts all the burden on individuals to figure out how their data is being exploited, the GDPR puts all the burden on the big data companies in terms of their accountability for the fate of online data. In the language of the European Parliament, the legislation "puts the citizen back in the driving seat" and turns Toomas Hendrik Ilves's notion of data integrity into law.

The legislation enshrines the so-called right-to-be-forgotten ruling that gives individuals the right to have their personal data erased from the internet. A person must provide "clear and affirmative consent" if data companies are going to process his or her private data. The law requires companies to inform individuals if their data has been hacked or tampered with, and it gives people the right to switch their personal data to another service provider. It also warns that companies face fines of up to 4 percent of their global revenue for breaking these new laws.

The importance of the GDPR legislation was reiterated to me by Paul Timmers, the director of Digital Society, Trust and Cyber Security at the European Commission, with whom I also spoke when in Brussels. What Timmers calls "digital" is now, in his language, "mainstreaming" in all EU policy areas, from energy and health care to transportation and education. The challenge, he said, is to use this new legislation to build the kind of trust regarding personal data that already exists in Estonia. If the GDPR legislation is properly enforced, Timmers added, it will also trigger enormous innovation around privacy. Trust, he reminded me, is the "most important currency" of our digital age. An entrepreneur who is able to build new companies based on what Margrethe Vestager called "privacy by design," he predicted, will be the next Jeff Bezos or Mark Zuckerberg, creating the next Amazon or Facebook of the digital age.

So what are we to make of all these new laws? Some of this regulation, such as the EU's determination to shine a light on how Facebook really uses its customers' data or the EU's efforts to make social media networks accountable for the fake news published on their platforms, is enormously valuable. In fact, after Commissioner Ansip's warning and the threat of fines from the German parliament, four American tech companies—Google, Facebook, Twitter, and Microsoft—all signed on to an EU "code of conduct" not only to take down illegal hate speech within twenty-four hours of publication but even to come up with "counternarratives" to fix the lies.[48]

Some of it, such as the Max Schrems case, reflects what Estonian president Toomas Hendrik Ilves aptly described as the "paranoid shit storm" over privacy that was triggered by Edward Snowden's NSA surveillance revelations. And some of it, quite frankly, is less than useful. The EU's Google tax, for example, is a particularly self-defeating project because it actually ends up harming publishers by reducing their web audience. In Spain, where a law was actually introduced for charging Google for using snippets, it forced Google to close its Google News service in the country and resulted in online publishers losing 10 to 15 percent of their web traffic in one dumb regulative stroke.[49]

The biggest issue of all, however, is whether all this regulation will realize Margrethe Vestager's goal of leveling the playing field for innovative EU tech entrepreneurs. As the *Financial Times* asked about the Schrems decision, "Will it help or hurt European start-ups?"[50] That's the multibillion-euro question, especially in the context of the radically disruptive GDPR regulations that will be launched in 2018.

So will all these new EU regulations fix or wreck the future? Is Europe on the brink of an innovation spring or a regulatory winter?

Innovating Big Brother

Renate Samson, the CEO of Big Brother Watch, a British civil liberties group focused on the defense of online privacy, had emailed me while I was Brussels. She knew I was writing a book about fixing the future and generously offered to help me with my research.

"There are some start-up entrepreneurs in London I think you should meet," she'd written, "so why don't you pop over for a chat?"

I'd met the effervescent Samson the year before at a transatlantic weekend retreat focused on the accountability of the private sector in the digital economy. The off-the-record event had been held at Ditchley, the eighteenth-century Georgian country estate near Oxford, where Winston Churchill spent much of the Second World War. The small, by-invitation-only event had been attended by public policy chiefs from such Silicon Valley companies as Apple and Twitter; senior US policy makers such as Julie Brill, a commissioner at the Federal Trade Commission (FTC), the US government agency supposed to regulate antitrust; and civic activists like Renate Samson. It had been an intellectually stimulating weekend, with what might be described diplomatically as a "frank exchange of views" between representatives of the government, private companies, and civic groups. Samson in particular had been openly critical of what she described as the "untrustworthy" business practices of private companies trying to "monetize" people's data.

I was intrigued as to why Renate Samson, the head of Big Brother Watch, a nonprofit group focused on the protection of online privacy against the prying eyes of both government and private enterprise, would want to introduce me to start-up tech entrepreneurs. After all, these insatiably data-hungry monetizers of our personal information were,

I'd assumed from what Samson had said at Ditchley, the very people from whom Big Brother Watch was supposed to be protecting us.

As a serial start-up entrepreneur myself, I know one thing for sure: There are no original ideas in tech. I'm certain that any "original" idea I come up with for a start-up will have been simultaneously "invented" by any number of other entrepreneurs around the globe. When the time is right for a new idea, it seems to show up in many people's minds at the same time with a serendipity that appears almost engineered by the gods. So every start-up guy in the mid-nineties (including myself)—and not just a brilliant young financial analyst with the idea of founding an online bookstore—had the same crazy dream of opening an e-commerce website. Ten years later, many smart geeks—and not just a particularly smart one in a Harvard dorm room—experienced that eureka moment of recognizing that the world needed an online social network. Today, with AI being all the rage in Silicon Valley, venture capitalist friends tell me that they listen to the same pitch for the same "my-smart-automation-start-up-is going-to-revolutionize-the-world" idea all day long.

So it is a little disconcerting when the first start-up team members to whom Samson introduces me explain, with breathless excitement, that they have the "first" kind of "enriched data" business. It's a start-up, they boast, that will "finally" empower individuals to be able to "own and control their own data." They are a couple of well-seasoned, middle-aged entrepreneurs—one a former Asia-based executive at a multinational media company and an angel investor, the other a partner at an international management consultancy and a former head of research at a big bank. But in their unabashed enthusiasm for their new product, they resemble excited teenagers.

These entrepreneurs tell me that the "best thing a government can do is provide the regulation that empowers people." Technology has become so cheap, they explained, that there's no need for huge public investment in new products and services. Once a law like the GDPR has been set, we can leave the rest to the private sector.

Their start-up has Margrethe Vestager's idea of "privacy by design" at its heart. It is a secure digital locker— built from supposedly "fail-safe" technology developed by an American defense firm—that allows people to securely store their personal online data. Rather than touching, seeing, or holding our personal information, this locker encrypts the data to ensure its security and portability. It had been designed as what its creators call the "plumbing" of a new, post-GDPR data economy in which privacy has become the new norm and trust its new currency.

So what's Big Brother's position on this? I ask Samson after the first set of entrepreneurs has left.

"Well, we've got to trust something," she says. Her point wasn't to fetishize the free market or the Randian entrepreneur, or any of the other libertarian orthodoxies that usually accompany these kinds of pitches for start-ups. That *something* we need to trust, Samson is saying, is a concrete law, guaranteed by the twenty-eight members of a community of five hundred million people, designed around the protection of data. That law, which turns the surveillance economy on its head, is doing Renate Samson's job for her. It might even make Big Brother Watch redundant.

To borrow some words from the former New York Yankees shortstop Yogi Berra, listening to the second entrepreneur, a self-funded former telecom executive, is "like déjà vu all over again." It feels as if we are back in 1995 and everything is once again possible. He even begins with the same promise of being the "first" kind of data company to create a product designed for "people." He then talks, with the same enthusiasm, about the importance of the GDPR in the creation of a privacy ecosystem, explaining that his start-up would've been unimaginable before this law had been passed. And he ends on the same note too—explaining that what's missing in the current big data economy is trust and that his product will reestablish trust at the center of the networked world.

I'm not sure if either of these businesses will succeed. They might, they might not. As a start-up entrepreneur, the other thing I know for sure is that predicting the future of other people's start-ups is the ultimate crapshoot. This was a pre-Brexit conversation—so by 2018 it's quite possible that these British entrepreneurs will have to relocate their GDPR start-ups to Brussels, Barcelona, Budapest, or Berlin. But the point is that there are entrepreneurs all over Europe—from Brussels to Barcelona to Budapest to Bristol—all waking up to the same innovative possibilities in a digital economy turned on its head by government legislation. Somebody will succeed at some point with a product that makes privacy the new norm. After all, as Renate Samsom says, nobody wants to be watched in everything he does.

Although so much of this reminded me of the landscape back in 1995, there is one critical difference. The first internet revolution was a purely free market affair. Today, however, we are already in Steve Case's Third Wave, when what he calls the "visible hand" of government will be "central" to the success of digital businesses. By 2018, then, it may have become self-evidently clear to everyone but the most die-hard libertarian that regulation is, indeed, innovation. That can only be good news for those of us, like Steve Case or Margrethe Vestager, who want to flatten the winner-take-all landscape and build a fairer world.

Not everyone, I'm guessing, will agree that the best way to fix the future is through legislation. There will be those who say that rather than Case's visible hand, it's actually what Adam Smith called the "invisible hand" of the free market that remains the best guarantor of innovation. Perhaps. To test this theory, let's return to where we began this journey—to the Alte Teppichfabrik, that industrial relic in Berlin, where some of Germany's most innovative free market digital entrepreneurs are attending the "Encrypted and Decentralized" conference.

CHAPTER SEVEN

COMPETITIVE INNOVATION

Re-Decentralization

Up on the top floor of the Alte Teppichfabrik, the old Berlin carpet factory with the chic industrial interior and the panoramic views over East Berlin, it's also 1995 all over again. In spite of Edward Snowden's rather apocalyptic warnings from Moscow, the "Encrypted and Decentralized" conference is abuzz with a new optimism about the old dream of digital reinvention. The hope is that the paired technologies enabling decentralization and encryption can provide us with what Snowden would call the "agency" to reengineer the internet economy. The vision, to borrow a term from the World Wide Web inventor Tim Berners-Lee, is the "re-decentralization" of economic power. It's an attempt to return to the original architecture of the web—to a network where power lies on the edge rather than in the center.

The conference invitation from BlueYard Capital had said, "We need to encode our values not just in writing but in the code and structure of the internet." Those "values," however, aren't just ethical—they are also the *values* of the free market. As the event's keynote speaker, Brad Burnham—the cofounder and managing partner at the New York City–based venture capital firm Union Square Ventures—tells the Berlin audience, "The great thing about capitalism is that it's the only choice."

And so, according to Burnham at least, it's the invisible hand of the market rather than the visible hand of Vestager's EU that can now reinvent the digital economy. "We are now living in a world resembling 1995," Burnham explains. "It's a world of smaller companies, a new age of innovation from below."

As I've already suggested, there really aren't any original ideas in tech. Today it's Berners-Lee's and Burnham's re-decentralization idea that is fast shaping the contemporary tech zeitgeist. To go forward, this idea is saying, we must return to the original principles of the web. To fix the future, we have to go back to 1995.

You'll remember there was even a parallel event in the same month as the Berlin conference—the "Decentralized Web Summit" at San Francisco's Internet Archive, featuring many of the internet's original architects, including the inventor Berners-Lee and the TCP/IP creator Vint Cerf. Everyone, it seems, on both sides of the Atlantic, is nostalgic for the future. "The web's creator looks to reinvent it," as the *New York Times* described this June 2016 event, which brought together privacy advocates and pioneers of such peer-to-peer technologies as blockchain to discuss a "new phase of the internet."[1]

Brewster Kahle, the Internet Archive founder and summit organizer, believes that the time is now right for a radical re-decentralization of digital power. The future has finally caught up with us, he tells me when I visit him at his office in San Francisco's Inner Richmond district. "Now is the time to finally create a decentralized web," he says, "by building values into the code itself." It's not a "trivial" task, he says, but it can be done.

"I asked Vint Cerf how difficult it was to build the original internet," Kahle tells me. "And Cerf replied, 'It took one year with a room of five or six guys.'"

The first time around, Kahle confesses, speaking about the original digital revolution, "we made it too hard for individual creators to get paid for their work." The mistake, he says, is that the web lost the ability to serve its users. "We can do better than this," he insists, "so that it isn't a security decision every time you click on a link . . . and so that all the online music and videos aren't owned by iTunes."

Many other digital pioneers share this vision of re-decentralizing the web. Ethan Zuckerman, the director of the MIT Center for Civic Media and another key member of the founding generation of internet geeks, believes that this struggle between the forces of centralization and decentralization has been a familiar feature of the digital economy even as far back as 1993, when there were no online directories for navigating the web. "For old-school cyberutopians like me, it is extremely disappointing that the internet is not inherently decentralizing," Zuckerman told me when I visited him at his office in Cambridge, Massachusetts, "but it's hard to build the opposite." Like many other idealists, he is critical of the entire ecosystem of today's online economy. He argues that we've become too dependent on the advertising business model that compounds the already exaggerated power of winner-take-all advertising companies like Google, YouTube, or Facebook. The more we publish, the more dominant these companies become, he says. So rather than technology, the fundamental problem with the web today is its dominant business model. The challenge, then, is the reinvention of internet economics. It requires us to rethink the whole ecosystem—from the free content and free services to the ubiquitous advertising and surveillance—of the digital economy.

The primary focus of the Berlin "Encrypted and Decentralized" conference was also the imagining of a better digital ecosystem for our networked age. But in contrast with Kahle's and Zuckerman's hopeful visions of what a new internet *might* look like, all the talk in Berlin is about a new digital world, a real betaland, that is *actually* coming into existence. Burnham's keynote address, "The Unraveling of Network Effects and Data Lock-Ins," is focused on the way in which the latest digital technology is turning the digital clock back to a more innovative age. "Network effects" and "data lock-ins" are the very forces that have transformed the open web of the mid-nineties into what Tim Berners-Lee bemoans as data "silos"—winner-take-all intermediaries like Amazon, YouTube, and Uber. And their "unraveling," according to Burnham, is happening because new technologies like blockchain and other "web protocols" are making these intermediaries—which he describes as "centralized data managing services"—redundant.

It's the end of the digital middleman, Burnham predicts; good-bye to the networked intermediary. One example of these protocols that Burnham cites in his speech is the open-source InterPlanetary File System (IPFS), designed to establish a permanent and decentralized method of storing and sharing files. Protocols like IPFS are allowing for the online exchange of data between independent players, resulting in the creation of what Burnham calls "decentralized marketplaces." Other examples of this are the so-called decentralized autonomous organizations (DAOs), such as the controversial peer-to-peer currencies Bitcoin and Etherium, which operate on blockchain technology. These networked platforms do away with the need for the middleman: a bank or a government agency. They are returning us to Berners-Lee's original web, a level playing field on which power resides on the edge, with its users.

After Burnham's speech, I talk with him outside the Alte Teppichfabrik in a small beer garden overlooking the river Spree. I begin with the same question that I'd asked Gary Reback about history repeating itself, first in 1995 and then now.

Raising his beer bottle, Burnham quotes Mark Twain. History doesn't exactly repeat itself, he answers, but it "rhymes" in a perpetual cycle of innovation. In 1995, he explains, the market opened up with the shift from the desktop computer industry, dominated by Microsoft, to a market defined by the web. The business model gradually moved from packaged software to open-source technology—a process that, in turn, gave rise to the Web 2.0 revolution of Google and Facebook. And today, he explains, we are back with a new kind of disruptive technology that is challenging the dominance of the winner-take-all companies of the Web 2.0 revolution.

Like Reback, Burnham—whose investments at Union Square Ventures include billion-dollar wins on Tumblr and Twitter—is a critic of monopolies. But in contrast with the antitrust lawyer's, Burnham's critique is a purely financial one. Venture capitalists, he explains, have a responsibility to their investors to ensure the biggest returns. Today's digital economy, with its tiny pool of winner-take-all companies, is actually as bad for venture capitalists as it is for everyone else. So the only way to improve the venture business is to figure out what he calls "mechanisms" for "opening up" the market again. In this more open future, the companies might be smaller, Burnham acknowledges, but there will be more winners and so the rewards for professional investors will be higher.

So how is this change going to come? I ask.

As opposed to Margrethe Vestager's or Gary Reback's, Burnham's faith remains in the free market to fix the problem of these dominant intermediaries. To him, the centralized services managing data are, ultimately, flawed businesses. He explains that things will become broken when Amazon or Uber increases what he calls "its take." So, in Uber's case, for example, the change will eventually be triggered by its drivers, who are getting more and more squeezed by the ridesharing company's increasingly onerous terms. Eventually, presumably, they will be pushed into participating in their own decentralized marketplaces that will compete with Uber. One possible platform for this type of radically new business is a German start-up called Slock.it that is building a blockchain-based "Universal Sharing Networks" which empowers people to bypass central silos like Uber and Airbnb and directly transform their cars into taxis or their houses into hotels.[2]

"But when?" I reply. When should we expect the emergence of that new ecosystem that will return online power to the user?

When? That, of course, is the billion-dollar question for a venture capitalist. It's all about timing. It's not even the business model of radical new companies like Slock.it that's critical. After all, Burnham confesses, he invested in Twitter when the short messaging platform had no idea how it would make money. Timing, then, is everything. In the venture capital business, it's always too early, until it's too late.

"It follows a familiar pattern." Burnham explains how massive change in the marketplace happens. "Nothing, nothing, nothing . . . And then something dramatic."

At "Encrypted and Decentralized," there isn't much evidence of the *something dramatic* that could change everything. There's certainly a lot of abstract technological talk about "distributed cloud systems," "blockchain databases," "DAOs," and "IPFSs," and how they all fit together into that ever-elusive "stack." But there's little evidence that any of this technology is actually maturing into an ecosystem of easy-to-use products and services that people really want. As so often at cutting-edge technology conferences, it is hard to separate the hype from reality. And all too often, I'm afraid, their only reality turns out to be hype.

Then, in the final discussion of the day, there is a panel on "sustainable ways of monetizing content." One of the panelists is Tim Schumacher, the cofounder and executive chairman of a Cologne-based company called Adblock Plus, an open-source app that prevents online advertising from showing up on your web browser. Schumacher is on the panel to discuss his company's new partnership with Flattr, a Swedish-based micro-donation provider. But before describing this deal with Flattr, Schumacher asks the audience for a show of hands about who is using Adblock Plus's product.

Everyone, absolutely everyone on the top floor of the old Berlin carpet factory raises a hand. They are all using the same disruptive technology that, in Brad Burnham's "age of innovation from below," empowers them to block online advertising—the very heart of today's conventional business ecosystem. So they are proving not only that the current business model of the internet doesn't work, but also that there might be an alternative ecosystem being reengineered in Germany out of its ruins.

Winning the Second Half

I arrange to meet Schumacher in Cologne's Cathedral Square, the civic space dwarfed by the city's Gothic Dom, Germany's most visited landmark and the largest church in northern Europe. Constructed predominantly between 1248 and 1473, the Cologne Cathedral, which boasts the largest facade of any church in the world, was finally completed only after the unification of Imperial Germany in 1880. This World Heritage site, therefore, is a testament not only to Germany's engineering heritage but also to the country's rich history in the reengineering of long-term collective projects.

This reengineering culture, while certainly not unique to Germany, is, nonetheless, a persistent historical feature of Europe's most populous and prosperous country. Germany's economic accomplishments, particularly in the late nineteenth century, were mostly based on its reengineering of the earlier, haphazard British industrial revolution with cutting-edge technology and planned investment in factories, infrastructure, and scientific research.

Today a similar kind of narrative seems to be shaping Germany's unfolding role in the world's digital economy. "Does Deutschland do digital?" asks the *Economist* about the world's leading engineering economy. In 1995, the response would certainly have been *no*; today, however, the answer is less clear. As the Deutsche Telekom CEO Timotheus Höttges admits, "The first half of the battle to master the digital world was lost." So the question now, Höttges says, is "How do we win the second half?"[3]

It's certainly an important question. The German business consultancy group Roland Berger estimates that the German economy will lose 220 billion euros of annual value if it fails to successfully transform itself digitally.[4] Today, only four of the world's 174 unicorns—privately held companies like Uber or Airbnb that have a valuation of more than a billion dollars—are German. Every key German manufacturing sector—particularly its 361-billion-euro automotive industry, which makes up 20 percent of total German industrial revenue and employs more than 750,000 workers—is now vulnerable to the revolution of the internet of things, with its billions of connected devices flooding onto the market each year. Fifty billion of these smart things by 2020, according to Cisco. And many more billions throughout the 2020s and 2030s. So what Germans call *Industrie 4.0*—the fourth stage of industrial evolution after water and steam power, mass production, and the information technology revolution—is of vital importance to the future of the world's leading engineering power. Even German chancellor Angela Merkel speaks regularly about the need to digitalize Germany's *Plattform-Kapitalismus*. "This period will determine the future strength of the world's leading industrial centers," Merkel said in April 2016, expressing her angst about winning the race to dominate the digital landscape. "We have to win this battle."[5]

So how can Germany win the second half of the digital game? The answer probably lies in the country's historical tradition of successfully reengineering technological revolutions begun elsewhere. As Christoph Keese, the author of *Silicon Germany*, notes, "Germany's great strength is evolutionary, incremental innovations." So even in the first half of the digital game, when Germany all but failed to get on the scoreboard, its most notable success lay in the replication of Silicon Valley innovation. The Berlin-based Rocket Internet, founded in 2007 by three brothers—Marc, Oliver, and Alexander Samwer—is an industrial-scale incubator of start-ups that clones models of American e-commerce companies throughout the world. Employing 30,000 people in more than 100 companies active in 110 countries, Rocket Internet has a $3 billion market value essentially based on taking other people's ideas and executing them effectively.

The problem with the Samwers' controversial Rocket Internet is that it's more of a clone factory than a true reengineering business. For German companies to win the second half, they have to do more than simply take other people's ideas and execute them effectively. The challenge now, as Brad Burnham argued at "Encrypted and Decentralized," is to be able to reengineer entire ecosystems. It's a challenge that—for the typical top-down German engineering firm that still imagines the economy being partitioned into stand-alone industrial silos—is quite daunting. The LinkedIn cofounder Reid Hoffman quipped that every twenty-first-century company is a technology company. And as a tech company, every company is now, like it or not, in competition with every other tech company. In the automotive industry, for example, one of the most pressing challenges for German car manufacturers is to determine where they will fit in the so-called software stack that will empower self-driving vehicles. In a new ecosystem driven, so to speak, by autonomous cars, their challenge is to avoid becoming the dumb commodified hardware at the bottom of the stack in an economy where, as you'll remember Marc Andreessen saying, software is eating the world. So for Mercedes and BMW, the most dangerous long-term competition now comes from Silicon Valley. Their existential threat is that they could be eaten by Google's, Tesla's, or Apple's algorithms, not by Toyota or Ford cars.

In 2013, to hoots of condescending laughter from the digital cognoscenti, Angela Merkel, using language she might have borrowed from More's *Utopia*, called the fifty-year-old internet *Neuland*—meaning "new land" or "uncharted territory." And yet in a sense, Merkel was right. If Germany is to win the second half of the game, the internet needs to once again become uncharted territory. All its orthodoxies need to be challenged. Just as the internet has disrupted old industries, so the time has now come, in the perpetual creative storm of our Schumpeterian capitalist economy, for outsiders to engineer the internet's own disruption.

Which brings us back to my meeting with Tim Schumacher in Cathedral Square. We meet in an Italian café adjacent to the cathedral. Sadly, this expansive Cologne square had recently become one of the most notorious places on earth. On the night of December 31, 2015, there was a riot in which groups of mostly immigrant men surrounded and then sexually assaulted women who were celebrating the incoming year. And so, along with the kids on bicycles, a flock of Asian tourists, and the yellow-T-shirted Hare Krishna people, there is a squadron of blue *Polizei* cars parked in the middle of the square, guaranteeing order on the cold, blustery spring afternoon.

Over tea, the modest, soft-spoken German entrepreneur tells me about a new kind of publishing ecosystem that is emerging on the internet. Built around disruptive platforms such as Adblock Plus, the Swedish content-sharing platform Flattr, and the new German search engine Cliqz, it reflects Brad Burnham's vision of a new age of innovation being created from below by small, nimble companies unencumbered by legacy business models or technologies.

A serial start-up guy who sold his first business, a domain names company, in the late nineties, Schumacher was looking around for a new project when, in 2010, he stumbled on the Adblock Plus software, which he describes as an "undiscovered gem." Originally created by a Moldovan engineer, the open-source technology was a volunteer-staffed, community-supported project without any business model. What it did, by hacking into browsers like Firefox or Chrome, was provide an advertising kill switch, replacing all adverts on websites with blank boxes. It turned the whole digital economy on its head. Rather than being perpetually watched and pestered by advertisers, the user now had the upper hand to block advertising, tracking, and cookie notifications, and even to disable those social media buttons that follow us around the internet. Adblock Plus enabled the destruction of the web's dominant business model with the click of a mouse.

Schumacher incorporated the company, raised 200,000 euros of seed money, and began to grow the business. In August 2011 it had fifteen million users. By March 2016 its user base had grown to one hundred million, making it by far the largest advertising-blocking software company in the world. The market is now growing by 50 percent a year, and Schumacher expects to have a billion users by 2020—giving him an almost Facebook-size community of users. Indeed, Adblock Plus has been so successful that even Google, the online advertising behemoth, has announced plans to build its own proprietary ad blocker into its Chrome browser in 2018.[6] Germany might not have had a great first half in the digital game, but with products like Adblock Plus being used by a billion people around the world, its second half is turning out to be a lot more promising.

Gazing out onto the spotless Cathedral Square, with its squadron of *Polizei* cars maintaining order, Schumacher tells me that, by allowing people to eliminate all the crappy advertisements that trash up so many web pages, he is helping to clean up the internet. Adblock Plus, then, can be thought of as a kind of digitally green solution to the digital pollution of our networked age. Schumacher tells me he is originally from Swabia, the southwestern region of Germany famous for its extreme cleanliness. What he's doing with Adblock Plus, the entrepreneur thus suggests half seriously, is reengineering the internet along Swabian lines.

Certainly there's a desperate need to clean up an online advertising industry that Schumacher describes as "lawless." Along with the disturbing surveillance issues associated with digital advertising, it's an industry increasingly corrupted by both false promises and outright fraud. "The digital age of transparency is just as often an age of obfuscation," complains the British author and communications expert Ian Leslie, bemoaning the complete lack of accountability of online advertising.[7] "Instead of perfect precision," Robert Thomson, the CEO of the global media empire News Corp, complains about what he calls the "digital duopoly" of Facebook and Google, "we have the cynical arbitraging of ambiguity" in which advertisers often end up unwittingly sponsoring hard-core porn, neofascist, and militant Islamist sites.[8]

A 2016 research report from the US Association of National Advertisers (ANA) and White Ops predicted that almost 10 percent ($7.2 billion) of the $77 billion spent on online display ads that year would be invested in fraudulent products.[9] Much of this fraud is made up of the use of bots imitating the browsing habits of real people, thereby creating an industry in which, according to one study by the

ANA, up to 37 percent of its supposed audience are actually zombies. Moreover, the online advertising industry—with its myriad layers of brokers and agencies and its byzantine systems for the automated buying and selling of ads—is, according to the *Financial Times'* business columnist John Gapper, "bafflingly complex." It's so complex, in fact, he says, that even "regulators are failing to block fraudulent ads."[10] So, rather than with the visible hand, the solution may lie in the invisible hand of the market, in the consumer demand for digital cleaning services such as Adblock Plus that, at the click of a mouse, banish the trash from our screens.

Schumacher is perfectly serious when he claims that Adblock Plus is establishing a healthy new publishing ecosystem for the internet. Many publishers, however, completely disagree—arguing that this ad-blocking product is actually destroying their advertising-supported businesses. Some—including *Forbes*, *Wired*, *Business Insider*, and, most notably, Facebook—don't even allow Adblock Plus users access to their content. Six big German publishers, led by Springer, are even suing Adblock Plus, claiming that the software modifies websites and thus should be illegal under copyright law.

When I talked to Mark Thompson, the *New York Times* CEO, he said he was "hyper-hostile to ad-blocking software." He described Adblock Plus as a "shit business," claiming that it's a new kind of protection racket that is "intent on exploitation." What Thompson so strongly objects to is Adblock Plus's creation of a business model that requires larger publishers to pay in order to be included in a white list of "acceptable ads." In Thompson's mind, the German company has become the internet's latest gatekeeper and is using its new power to demand rent from large publishers like the *New York Times*. The disruptive success of Adblock Plus is certainly forcing Thompson to rethink his business. In June 2016 he confirmed that the *Times* would launch an advertising-free digital access option that would be priced higher than the regular subscription.[11] But ad-blocking technology doesn't seem to have hurt the *New York Times* business model of selling subscriptions of its high-quality content. Since the election of Donald Trump in November 2016, the rate of sale of *Times* subscriptions is up more than tenfold, raising overall subscriptions to 2.5 million subscribers, adding more than $30 million a year, and now meaning that subscriptions make up more than 60 percent of the company's revenue.[12]

Schumacher, however, argues that Adblock Plus is acting as a "trustee of the consumer" in empowering users to configure the platform. "We are establishing rules," he insists, "on a lawless industry." And, along with these new rules, he's also trying to pioneer a new online publishing business ecosystem in his partnership with the Swedish micropayments site Flattr, a decentralized marketplace that brings together producers and consumers of content. The plan is to combine Adblock Plus with Flattr's 200,000 users and 30,000 publishers—thereby taking out the intermediary and enabling these consumers to directly fund advertising-free online content, particularly journalism. No, it's not Mark Thompson's iconic *New York Times*, with its fabulously enlightening content created by professional journalists and curated by teams of editors and fact-checkers. But it is a promising example of Brad Burnham's faith in the invisible hand of the free market to create digital innovation from below.

Flattr was founded by Peter Sunde, the Swedish digital activist who had also cofounded the controversial network The Pirate Bay—the index of digital content operating on the BitTorrent peer-to-peer protocol that was, in truth, mostly a platform for the exchange of stolen goods. Sunde, Schumacher promises me with a sad smile, brings "the passion of the convict" to this partnership with Adblock Plus. This is no exaggeration. Sunde was imprisoned for a year in Sweden for allegedly assisting Pirate Bay users with infringement of copyright. So his partnership with Adblock Plus is an attempt to reengineer a legal ecosystem for an online content industry that, in a previous life, he helped undermine.

In my 2007 polemic, *The Cult of the Amateur*, I was extremely critical of the way in which peer-to-peer networks like The Pirate Bay were enabling the theft of online content and thus destroying the livelihood of musicians, photographers, writers, and filmmakers. And so, over the years, Sunde and I had often been on opposite sides of the piracy debate, sometimes even personally clashing in discussions about the need for copyright law and the (im)morality of online piracy.

But when I meet Peter Sunde for dinner in Copenhagen to discuss Flattr and his partnership with Adblock Plus, I discover that, while he was busy trying to reinvent the internet, he had also reinvented himself. No longer interested in even discussing the rights and wrongs of The Pirate Bay, Sunde tells me that his focus now is on rewarding creative people for their work and creating decentralized alternatives to Google and YouTube.

"We need sustainable funding for creative people," he says, "to ensure that creative people are paid."

Peter Sunde has learned from his youthful mistakes. Rather than participating in the killing of our culture, he is now working on its rebirth. It might have taken him a while to arrive at More's Law, but in his attempt to reinvent both himself and the internet, he has become a valuable player on Team Human.

Unsafe at Any Speed

At Europe's most prestigious tech gathering, the Digital Life Design (DLD) Conference in Munich—that same event, you'll remember, where the EU antitrust chief Margrethe Vestager confessed to being unconcerned about the fragmentation of the internet—there was speculation about Germany's chances of winning the second half of the *Industrie 4.0* game to control today's internet of things. The discussion focused on how local players could barge their way onto the global *Plattform-Kapitalismus*.

But many of the American attendees at DLD, particularly those from Silicon Valley, remained pessimistic about Germany's ability to turn around its poor first-half performance. One investor from Sand Hill Road, the Palo Alto street adjacent to Stanford University that has become a synonym for venture capital firms, told me that Germany lacked the angel investment networks and the cultural acceptance of failure to "do digital" successfully. Others complained about the conservative nature of German corporations, particularly the absence of a risk-taking ethic among senior executives. Above all, however, their critique focused on the local culture—Germans' tendency to look backward, to be focused on their complicated history rather than on the future, many Americans at DLD observed. Germans remain more preoccupied, the Sand Hill venture capitalist quipped, with fixing their past than with fixing their future.

Others were even less charitable. "The Germans make great cars," one brash young founder of a San Francisco–based AI start-up told me over lunch in Munich. "That's what they are good at. They just don't do start-ups."

Some of this, of course, is true. Compared with Silicon Valley, German business culture is quite conservative. But what many of the Americans at DLD overlooked, especially the younger and brasher start-up entrepreneurs, is that a historical perspective on the future often not only is wise, but also has significant economic value.

The Union Square Ventures cofounder Brad Burnham, you'll remember, argued that we are in a "new" age of innovation that resembles 1995. But there is another twentieth-century year that might be equally valuable in terms of making sense of our twenty-first-century future. That year is 1965.

It was in 1965, of course, that Gordon Moore invented his eponymous law in an article titled "Cramming More Components onto Integrated Circuits," first published in a special *Electronics* magazine issue dedicated to predicting the future.[13] But in 1965 there wasn't much popular interest in technical white papers about integrated circuits. Back then, anyone interested in reading about new technology had bought a book that not only turned out to be the year's most influential work of nonfiction, but also changed an entire global industry. Written by Ralph Nader, the book was called *Unsafe at Any Speed: The Designed-In Dangers of the American Automobile*, and just as Rachel Carson's 1962 bestseller *Silent Spring* dramatically raised public consciousness about the dangers of pesticides and toxic chemicals in our foods, so Nader's 1965 book did the same thing for the deadly dangers of American automobiles.

The most interesting presentation at DLD was by a young German entrepreneur called Marc Al-Hames, the co-CEO of Cliqz, the new internet browser and search engine. Al-Hames's speech, borrowing from Nader's 1965 bestseller, was titled "Unsafe at Any Speed." The first slide Al-Hames showed to the DLD audience was of a brand-new two-door convertible red Corvair, a car manufactured by the American automobile company Chevrolet between 1960 and 1969.

"This was the car you wanted to have in 1965. It had a huge engine, incredible acceleration, lots of chrome. But there is only one problem about this car," Al-Hames explained, referring to the gleaming Corvair, which was designed to be a combination of the Chevrolet's iconic Corvette and Bel Air models.

"That problem," Al-Hames explained, "is that it was unsafe at any speed."

In the mid-fifties, the big three American automakers, Ford, GM, and Chrysler, controlled 96 percent of the American market. As competition grew more intense, these carmakers responded to a public appetite for automotive novelty, gadgetry, and space-age design with meretricious vehicles like the Chevrolet Corvair, which were designed to catch the eye and then wear out quickly. The goal, according to one industry executive, was to get consumers to buy a new car each year. Safety concerns became, at best, an afterthought, and these seat-beltless cars became little more than curvaceous chrome-plated coffins. In 1961, the annual number of traffic fatalities in America was 38,000. In 1966 it had soared to 53,000, a shocking 38 percent increase in just five years.

Then Nader, a young attorney at the time, published *Unsafe at Any Speed*, the most influential book ever written about car safety. "A great problem of contemporary life," Nader wrote, "is how to control the power of economic interests which ignore the harmful effects of their applied science and technology." The point of his book was to control those interests.

His first chapter, "The Sporty Corvair: The One-Car Accident," exposed the Corvair's intrinsic suspension problem, which made the Chevrolet so incredibly dangerous to drive. The book detailed the gruesome flaws of the Corvair's design—such as its death trap of a steering wheel, which failed to collapse on impact, thereby impaling its drivers and subjecting them to the most gruesome death or disfigurement in an accident. And it accused Chevrolet of deliberately ignoring the connection between the car's ostentatious design and its absence of safety features. But the book was also a broader wake-up call about "highway carnage" that in 1964, Nader argues, was costing America an annual $8.3 billion (the equivalent of more than $66 billion today) in property damage, medical expenses, lost wages, and insurance overhead expenses.[14] *Unsafe at Any Speed* was a public relations catastrophe for an American car industry that, in many ways, is only now recovering its mojo with Elon Musk's electric car start-up Tesla—the innovative Silicon Valley–based company that, in April 2017, passed General Motors to become America's most valuable automobile manufacturer, with a $52.7 billion market cap.

Like the American food industry, the car market over the half century since Nader published his exposé has been dramatically reshaped by our five-pronged forces of government regulation, competitive innovation, social responsibility by citizens, worker and consumer choice, and education. Nader's role, as a concerned citizen educating the public about the dangers of the Corvair, represented the beginning of a fifty-year story that has resulted in a dramatic improvement of car safety in America. In 1965, when *Unsafe at Any Speed* was published, there were five deaths per hundred million miles traveled. By 2014 this number had dropped to one death[15]—an astonishing 80 percent decrease in driving fatalities.

How did this happen? Just as government regulation literally cleaned up the US food industry, the same was true of the automotive industry. Nader's book led, in 1966, to the creation of the Highway Safety Act and the National Traffic and Motor Vehicle Safety Act. That same year, the Department of Transportation was created. These new government acts and the department set new safety standards, such as 1966 legislation mandating that all cars should have seat belts. Subsequent legislation led to seventeen other safety features, including requiring new vehicles to have padded seats and dashboards and improving the locks on car doors. Between 1966 and the late 1980s, more than eighty-six million cars were recalled under the new federal rules. After New York and other states enacted their own seat belt laws in the 1980s, seat belt usage rose from between 3 and 10 percent in the '70s to 73 percent by 1994.

Unfortunately, at least for Detroit, it was German car manufacturers who reinvented their products and marketing to satisfy a new consumer demand for safer vehicles. Daimler-Benz's safety cage, patented in 1951, offered a front and rear "crumple zone" that became a feature in Mercedes cars by 1959. Other innovations German manufacturers added to their cars after 1966 included windshields with safety glass, "burst-proof" locks for doors, shock-absorbent bumpers, anti-whiplash head restraints, and a collapsible steering column. In 1967 Volkswagen debuted electronic fuel injection, which dramatically lowered toxic emissions and reduced fuel consumption. In 1970 Daimler-Benz unveiled antilock braking on its cars. The overall result was a stunning shift in consumption patterns. From that 96 percent share of the domestic car market in the mid-fifties, the big three of Chrysler (13.2 percent), Ford (15.6 percent), and General Motors (17.1 percent) made up just over 45 percent of the US market in 2017. Meanwhile, Americans are buying so many luxury German cars today that President Trump has threatened to slap a 35 percent tariff on the "millions of cars," he claims, that are flooding into the United States from Germany.

So let's fast-forward half a century. What the hell, you are probably wondering, does a book about the 1965 Corvair have to do with a contemporary German internet browser and search engine? The connection, according to Al-Hames's DLD speech, is that both industries—the

American automobile industry and the American internet industry—are "unsafe at any speed." Of course today's internet technologies aren't literally impaling their users. And yet, according to Al-Hames, Silicon Valley is at a stage similar to that of the car industry of the mid-sixties in terms of losing both the trust and the loyalty of its consumers. Echoing the arguments of digital pioneers like Tim Berners-Lee and Brewster Kahle, Al-Hames argues that only 50 percent of web pages are secure, that trackers spy on us on every web page—with 25 percent of the pages having more than ten spies—that none of us have any idea where our data is going, and that 49 percent of users don't trust the internet. Like the Chevrolet Corvair, Al-Hames argues, the current internet, with its surveillance-style advertising-based ecosystem, "isn't sustainable."

This is what has led to the founding in 2015 of Cliqz, one of the most ambitious and well-financed digital media start-ups to have recently come out of Europe. Backed with tens of millions of euros in early-stage investment from Hubert Burda Media, the third-largest media company in Germany and the organizer of the DLD Conference, Cliqz already employs more than a hundred people, even though the company currently has no revenue model. Paul-Bernhard Kallen, the innovative CEO of Burda, explained to me why Burda has invested so much time and money in Cliqz. "Something seems to have gone wrong" with the internet, he explained when we met at DLD. Like so many other critics of today's networked society, Kallen believes the central problem is the disappearance of trust, caused mostly, in his opinion, by "too many people connecting too much data." Burda, he explained, has a moral responsibility not just to make profits but also to figure out ways that we can once against trust media. To do this, Kallen explains, we need to "rethink" and "reinvent" search, the heart of the information economy.

Created by Jean-Paul Schmetz, the former chief scientist at Burda, Cliqz is an innovative combination of search engine and internet browser, built on "privacy by design" principles. It's been engineered or, perhaps more exactly, *reengineered*, as the anti-Google—an internet browser with an inbuilt search engine designed in such a way that it will never collect or sell its users' data. But unlike the digital locker products introduced to me in London by the associates of Renate Samson, Cliqz hasn't been created to take advantage of any new EU data legislation. The German start-up believes that it can win purely through consumers choosing this superior product—that is to say, through the invisible hand of the market. Cliqz is ready to directly take on Silicon Valley leviathans such as Google because Marc Al-Hames and Jean-Paul Schmetz believe their product is the best on the market, at least for a privacy-concerned user—which, in truth, means everyone inside and outside Germany. Indeed, given that President Trump signed an April 2017 bill repealing laws protecting the privacy of internet users, there may ultimately be more demand for a product like Cliqz in the United States than there is in Europe.

Denelle Dixon-Thayer, the chief legal and business officer at the internet's third most popular internet browser, Mozilla (after Microsoft's Internet Explorer and Google's Chrome), certainly thinks so. "We love what they are doing, love that they are looking at alternatives for search," she told me when we spoke about Mozilla's strategic partnership with Cliqz.

"The business model of the web isn't broken," Dixon-Thayer adds. "We just need more transparency and honesty."

Which is where Cliqz comes in. "We might not have invented the internet," Al-Hames concludes his DLD speech, sounding uncannily similar to Adblock Plus's Tim Schumacher, "but we clean it up and make it a better place every day."

A few weeks later, when we meet for breakfast in Munich, I ask Al-Hames the same question I'd asked Brad Burnham and Gary Reback about history repeating itself. Only this time I substitute 1965 for 1995.

Al-Hames, who is as energetic as he is thoughtful, nods vigorously at the similarities between 1965 and today. Silicon Valley is not remotely prepared for the future, he explains. Like those shortsighted Chevrolet executives in the 1960s who believed they could continue forever to sell chrome-plated coffins to unsuspecting consumers, American tech companies just take it for granted that the current data ecosystem will remain intact. But the current system of tracking users, which has transformed the internet into a giant inspection house, he insists, is "out of control" and "eventually must change." Consumers simply don't want it. And just as Brad Burnham forecast a consumer rebellion against winner-take-all companies like Uber or Amazon that increase "their take," so, Al-Hames believes, we will eventually see a consumer rebellion against Silicon Valley's increasingly exploitative big data companies. That's what history teaches us about the future, he says. It's always a similar story with the same kind of ending.

In 1965, Al-Hames explains, who could have imagined a world where seat belts were legally required, where cars were designed around increasingly sophisticated safety features,[16] and where the number of deaths in the United States per hundred million miles traveled had been reduced by 80 percent? And back in 1965, who could have imagined that Volkswagen, which had only just begun exporting its consumer-friendly Beetle into the US market, would now have a billion-dollar production facility in Tennessee? Or that Chrysler, one of the big three automakers in 1965, would have filed for bankruptcy in 2009 and would now be owned by the troubled Italian car manufacturer Fiat?

Future generations, Al-Hames says, will look back at our own time in similar disbelief. It will be the same way we look back at 1965 and are shocked by the vast majority of people who didn't wear a seat belt while driving. In fifty years' time, he predicts, future generations will be equally horrified and amazed by the cavalier way in which our most personal information was revealed to unaccountable, opaque multinational corporations headquartered on other continents.

The data supports Al-Hames's argument. According to the Pew Research Center, 86 percent of American internet users have taken steps to disguise their digital footprints, and 91 percent of them agree that consumers have lost control of the way their personal information is used by companies.[17] As with the big three American automakers in 1965, it's increasingly clear that today's Big Data internet companies are unsafe at any speed.

"Nothing, nothing, nothing . . . And then something dramatic," is how, you'll remember, Union Square Ventures's Brad Burnham describes significant economic or technological change. It happened with food and cars, and it will eventually be triggered in the data economy too by a combination of innovators like Marc Al-Hames and regulators like Margrethe Vestager. As Mozilla's Denelle Dixon-Thayer reminds us, surveillance ultimately isn't a good business model. And if there's one thing history teaches us, it's that bad business models eventually die.

Chapter Eight

SOCIAL RESPONSIBILITY

A View from Nowhen

Sometimes the future shows up in the oldest of places. I'm having lunch with Huw Price, the Bertrand Russell Professor of Philosophy at Cambridge University and the cofounder, with Jaan Tallinn and Martin Rees, of Cambridge's Centre for the Study of Existential Risk. We are sitting in Trinity Hall, the seventeenth-century communal dining room of the Cambridge college that was founded by the Tudor monarch Henry VIII—the absolutist ruler who ordered the execution of his former chancellor Thomas More after More refused to countenance the legality of Henry's divorce from his first wife, Catherine of Aragon.

"It's the most famous and wealthiest college in Cambridge," Price tells me, not without a hint of pride about Trinity, where he is a fellow.

Over the last five hundred years Trinity has been one the world's most exclusive and powerful clubs—educating generation after generation of elites, including the inventor of the theory of gravity, Sir Isaac Newton; the father of empiricism, Francis Bacon; thirty-one Nobel Prize–winning scientists; British monarchs and prime ministers; and even the current Singapore prime minister, Lee Hsien Loong, who was awarded a mathematics and computer science degree from Trinity in 1974.

Price, a gangly, down-to-earth Australian whose only ornamentation is an expensive-looking timepiece wrapped around his wrist, is a most gracious host. As we eat our fresh poached salmon and salad in the historic Hall, we talk about the potential dangers of artificial intelligence and our moral duty to manage this new technology responsibly.

The Hall is an odd place for thinking about the future. Sitting opposite each other on antique pews in the cavernous dining room, we are enveloped by the past. It is a dizzyingly historical picture. Behind us hangs a late-sixteenth-century copy of Hans Holbein's life-size painting of a swaggering, vainglorious Henry VIII—all barrel chest, spread legs, sinuous calves, and clenched fists—which was originally completed in 1537 to celebrate the birth of the king's long-awaited first son, Edward. Holbein is the same Renaissance artist, you'll remember, who not only painted the best-known portrait of Sir Thomas More but also probably drew the map of Utopia for More's little book.

Just as the Hall is an incongruous place to be contemplating the future, so Huw Price, who first met Jaan Tallinn at a conference about the science of time on a Baltic cruise ship, is a strange person to ask about it. That's because, as a philosopher, at least, Price is skeptical about the very idea of time. Borrowing from Albert Einstein's general theory of relativity, Price is a proponent of the "block universe" theory, which suggests that time itself might be what he calls an "anthropocentric" idea.[1] "Physics has no sense of past or future," Price argues, and so time—that seemingly inevitable stream of moments, we intuitively assume, that always links yesterday and today with tomorrow—is an illusion.[2]

"There's no difference between the past, the present, and the future," Price explains. "It's a mistake to think of the future as open."

As I chew on my poached salmon, I nod politely. If time doesn't exist, I wonder to myself, then why is he wearing such a fancy-looking wristwatch?

In all seriousness, however, the block universe idea might not be quite as scientifically far-fetched as it sounds. Price quotes Albert Einstein's letter to the family of a friend who had just died. "For those of us who believe in physics," Einstein consoled the grieving family, 'the distinction between past, present and future is only a stubbornly persistent illusion.'"

Whether or not it's scientifically conceivable, this theory certainly has the most deliciously surreal implications. In his best-known philosophical treatise, *Time's Arrow and Archimedes' Point*, Price suggests that time's arrow—the direction of time—could just as easily be moving backward as forward. He thus presents the seemingly absurd idea of "backward causation" as a perfectly logical scientific truth. If Price's arguments are correct and time does, indeed, travel backward, then he and I—by eating lunch together in what I wistfully described as the "metaphysical" Hall, with its "hallowed" air—might, in fact, be the "historic" ones influencing the Reformation, the Renaissance, and the Enlightenment. If time's arrow really is directed backward, then nostalgia for tomorrow would be entirely logical. Jaron Lanier certainly wouldn't be the only person who missed the future.

Just as the classical Greek mathematician Archimedes called for a vantage point in the universe, the so-called Archimedes point, to be able to objectively examine the physical world—a "view from nowhere" as one contemporary philosopher put it—so Price believes that physicists and philosophers need to establish a similarly authoritative place, a point outside time, where we can think about the temporal world. He describes this place—where there is no yesterday, today, or tomorrow—as a "view from nowhen."[3]

But Price himself, at least in his life outside of philosophy, is no closer to this view from nowhen than any other temporally constrained person. He divides his own scarce time equally between his Cambridge professorship—the "best job in the world," he confides—and his work trying to fix the future. As the academic director of both the Centre for the Study of Existential Risk and another prominent new Cambridge institute, the Leverhulme Centre for the Future of Intelligence, Price is a key player on Team Human. The purpose of his two research institutes is to investigate the future before it happens. Price's institutes are hiring some of our best scientific researchers to determine the impact of new technology, particularly artificial intelligence, on society.

"Why worry about the future?" I ask. "What motivates you to make the world a better place?"

It's time, in fact, that triggered his concerns about the future, time that launched his own mission to make the world a better place. In 2011, Price explains, he became a first-time grandparent, of a baby girl born in Sydney, Australia. "For the first time in my life," he confesses, "I realized that there were people I cared about who would be around at the end of the twenty-first century."

Price tells me that he wanted to do something "useful" and "practical" with part of his life. In his role as director of these research centers, he sees himself as a start-up kind of guy, a facilitator of talent, what he calls "a builder of human networks for the twenty-first century." This pragmatic approach is his way of trying to change things, of doing his duty, of leaving the world in better shape for his granddaughter. It's his version of More's Law. Like Jaan Tallinn, he believes that smart machines are a potential menace—what the Australian philosopher colorfully describes as "a whole new demon"—to our species. He worries about the massive amount of money—$6 billion alone between 2014 and 2016—pouring into artificial intelligence start-ups, particularly given that small advances in the technology can create billions of dollars of value. But philosophy, particularly ethics, is never far from Price's mind. So he tells me that it's "very important" for venture capitalists to use moral criteria to determine their investments in the AI space. These kinds of ethical human judgments are critical, he says, if we are to maintain our control over smart machines.

Yet for all his concerns about the demonic potential of artificial intelligence, Price isn't entirely pessimistic about the future. He is encouraged, for example, by what he describes as the ethical maturity of the three cofounders of

DeepMind, particularly Demis Hassabis, its young Cambridge-educated CEO. This is the London-based tech company whose investors include Jaan Tallinn and Elon Musk, a start-up founded in 2011 and then acquired by Google for $500 million in 2014. DeepMind made the headlines in March 2016 when AlphaGo, its specially designed algorithm, defeated a South Korean world champion Go player in this 5,500-year-old Chinese board game, the oldest and one of the most complex games ever invented by humans. But in addition to the commercial development of artificial intelligence, Price explains, the DeepMind founders—with other Big Tech companies like Microsoft, Facebook, IBM, and Amazon—are helping engineer an industrywide moral code about smart technology.

This self-policing initiative, known, rather awkwardly, as the Partnership on Artificial Intelligence to Benefit People and Society, was formally launched in September 2016. Its goal is to make the world a better place. Trust us, the companies in this alliance say, promising a laundry list of feel-good issues, including "ethics, fairness, and inclusivity; transparency, privacy, and interoperability; collaboration between people and AI systems; and the trustworthiness, reliability, and robustness of the technology."[4]

Trust us with your future, they are saying. *Trust us* is, indeed, becoming a familiar promise from the tech community. The self-policing strategy of the DeepMind coalition sounds similar to the goals of another idealistic Elon Musk start-up—OpenAI, a Silicon Valley–based nonprofit research company focused on the promotion of an open-source platform for artificial intelligence technology. Musk cofounded

OpenAI with Sam Altman, the thirty-one-year-old CEO of
Y Combinator, Silicon Valley's most successful seed investment fund. Launched in 2015 with a billion dollars raised
by Silicon Valley royalty including the multibillionaires Reid
Hoffman and Peter Thiel, the Silicon Valley–based OpenAI
is run by a former Google expert on machine learning and
staffed with an all-star team of computer scientists cherry-
picked from top Big Tech firms.

"What a time to be alive!" Altman says about a networked age in which what he calls "the merge" of humans
and smart machines is actually happening. Even our data-
rich smartphones, he warns of this imminent Singularity,
"already control us." We are thus faced with a stark existential choice, he believes. "Any version without a merge
will have conflict . . . We enslave the A.I. or it enslaves us.
The full-on-crazy version of the merge is we get our brains
uploaded into the cloud. I'd love that. We need to level up
humans, because our descendants will either conquer the
galaxy or extinguish consciousness in the universe forever,"
he says, presenting this superintelligence threat as if it's the
plotline of a *Star Trek* episode.[5]

I ask Price what these young entrepreneurs, fabulously
wealthy and gifted technologists like Deep Mind's Demis
Hassabis, or Y Combinator's Sam Altman, need to incorpo-
rate into their self-prescribed moral code. What, I wonder,
should the new men of the twenty-first century be thinking about to ensure that his Australian granddaughter will
actually get to see the dawn of the twenty-second century?

From where are these "moral criteria" going to come?
I want to know. How can we be confident that these new

masters of the universe are actually working on behalf of *people*, with the interests of *society* in mind?

For a moment, Price reverses his identity, shifting from start-up guy back to being the Bertrand Russell Professor of Philosophy at Cambridge University. He reminds me of the core ethical principle of the Enlightenment, the "categorical imperative" of the eighteenth-century Prussian philosopher Immanuel Kant—a man of such regular temporal habits, it is said, that the people of the city of Königsberg would set their watches by his lunchtime stroll.

"Kant reminds us that with agency comes responsibility," Price explains.

Kant's association of agency and responsibility is the Enlightenment version of More's Law and the moral companion to Stephen Wolfram's and Ada Lovelace's definition of what it means to be human. Having agency, possessing goals, originating things means also being able to intuitively distinguish between right and wrong. And so, being human, according to Kant, in spite of what you'll remember he called humanity's crookedness, means naturally doing good, irrespective of who benefits from these actions. And although he's too self-effacing to make this point, Huw Price himself, by choosing to run his two Cambridge institutes in the public interest, could be a living example of this Kantian model of the responsible citizen.

Super-Citizens

It would, of course, be wonderful if everyone in Big Tech were a moral philosopher like Huw Price. Indeed, if that were the case, there would be no need for this book, since the philosopher kings of Silicon Valley would—civic algorithm by civic algorithm—be automatically fixing the future. Such an idea was the premise of the first and most influential of all imaginary societies, Plato's *Republic*. And it's been the subject of many of them ever since, including More's Utopia, the imaginary island that Hans Holbein represented as a skull.

"*Memento mori . . . Respice post te. Hominem te esse memento*," the Roman slaves would yell at the victorious general when he was returning from battle. "Yes, you will die, but until then, remember you are a man."

In a sense, Huw Price's block universe theory of time is correct. Nothing ever really changes. More's Law—with its insistence on our duty to our community—was as relevant in classical antiquity, sixteenth-century England, and nineteenth-century America as it is today. The challenge, as it's always been, is convincing the kind of elites who attend Trinity College that with great power comes equally great responsibility.

This is a particular problem in Silicon Valley, the center of power in our networked age. As I write this, in late June 2017, the tech community is once again embroiled in a series of scandals over its disgraceful treatment of women, with two high-profile investors, Dave McClure of 500 Startups and Justin Caldbeck of Binary Capital, forced to step down because of their persistent and well-documented attempts to sexually assault female entrepreneurs. Earlier this month, Travis Kalanick, the cofounder and CEO of Uber, was forced to resign because of a long history of ethical controversies at the ridesharing company in everything from rampant sexual harassment of its female engineers to threatening tech journalists to spying on its customers.

"What is most interesting is that Silicon Valley remains in a cognitive bubble, reluctant to engage with legitimate public worries over monopoly, privacy and tech related job disruption, not to mention its own culture," the *Financial Times*' Rana Foroohar wrote in July 2017 in response to this latest round of Silicon Valley scandals.[6]

So, yes, it's certainly not bad news that three of the Four Horsemen of the Apocalypse—Facebook, Amazon, and Google—have joined a self-policing alliance of Big Tech companies committed to "ethics, fairness, and inclusivity" in the development of AI products. But how much self-imposed morality can we really expect from multi–hundred-billion-dollar companies that are currently accused of a raft of unethical behavior around the world, including monopolistic market practices, exploitation of their users' data, and failure to pay their local taxes—not to mention the myriad other ethical scandals perpetually engulfing Big Tech? Although these companies are by no means intrinsically bad, they remain profit-driven businesses, answerable only to their shareholders or investors. Whatever seductive promises their *don't be evil*–style slogans make, there is no such thing as a moral for-profit company, inside or outside Silicon Valley. For better or worse, the aim of these private superpowers is to dominate markets, not share them. Their goal is their bottom line, not ethics.

And yes, it's also encouraging that powerful Silicon Valley investors like Elon Musk, Reid Hoffman, and Peter Thiel have contributed significant capital to an open-source AI platform that, we are promised, won't be owned or operated by a single data silo. But, I'm afraid, there aren't too many people in Silicon Valley quite as responsible as LinkedIn co-founder Reid Hoffman, a relative paragon of civic virtue, who, during the 2016 American presidential election, promised to donate five million dollars of his own money to a veterans' charity if Donald Trump publicly disclosed his taxes.

In spite of being an investor in OpenAI, Reid Hoffman is skeptical of hubristic Silicon Valley companies that believe they can stand outside history and fix the entire world. Rather than being courageous, he suggests, that's just myopic. Even juvenile. "It's great they are being ambitious," Hoffman told the *New Yorker* about Sam Altman and some of his Y Combinator projects. "But classically in the Valley, when people try to reinvent an area, it ends badly."[7]

Hoffman's ambivalence about the grandiose promises of OpenAI may be one reason why he is also one of the major investors in the "Fund for Artificial Intelligence and Society" launched by the nonprofit Knight Foundation in early 2017. This fund, which also includes the MIT Media Lab and the Berkman Center at Harvard as partners, and, you'll remember, Betaworks' John Borthwick as an advisor, is designed—like the Centre for the Study of Existential Risk at Cambridge—to foster a combinatorial network of researchers, ethicists, and technologists focused on study-ing the impact of AI on society. In contrast with the Deep Mind or OpenAI coalition, this Knight Foundation initiative doesn't just rely on technologists to make ethical decisions.

"My point of view is that it is a massive transformation and does really impact the future of humanity," Hoffman says about the AI revolution. "But that we can steer it more toward utopia rather than dystopia with intelligence and diligence."[8]

John Bracken, a longtime nonprofit executive who runs the Knight Foundation program, told me that Hoffman is a "particularly important" influence on this new fund. "We *trust* Reid Hoffman," Bracken, who has worked with Hoffman in the past, said to me. We trust him to know that with agency comes responsibility.

So what about the ethics of Big Tech's other multi-billionaires—the Cooks, Zuckerbergs, Benioffs, Bezoses, Pages, and Brins? None, of course, are as immature as Travis Kalanick. Nor are they as Kantian as Reid Hoffman. Their morality is, instead, a complicated question. Sometimes we can trust these plutocrats, sometimes we can't. Some, like Amazon founder and CEO Jeff Bezos, who at the time of writing has a net worth of $83 billion, making him the second-richest person on earth after Bill Gates, are simultaneously cutthroat businessmen and selfless philanthropists. On the one hand, Bezos is the much-feared leader of a semi-monopolistic e-commerce company that has highly questionable labor practices and a well-documented history of bullying publishers;[9] on the other, he's the public-spirited owner of the venerable *Washington Post* and—under a Trump regime that has castigated journalists as "enemies of the people"—among America's most powerful defenders of both a free press and democracy.

In June 2017, Bezos tweeted a "request for ideas" for philanthropy. "I'm thinking about a philanthropy strategy that is the opposite of how I mostly spend my time—working on the long term. For philanthropy, I find I'm drawn to the other end of the spectrum: the right now," he hinted about his philanthropic plans. "If you have any ideas," Bezos concluded, with typical quirkiness, "just reply to this tweet . . ."[10]

To take another example, Marc Benioff—the swaggering Henry VIII look-alike CEO of the business software provider Salesforce.com—though a remarkably generous supporter of many noble social causes, remains perhaps a little too enamored with attaching his name to public projects such as the UCSF Benioff Children's Hospital in San Francisco. Then there's Tim Cook, the Apple CEO, whom we last met in Margrethe Vestager's Brussels office, trying to convince the EU commissioner that the 0.005 percent tax his company paid to the Irish government was somehow in the public interest. Yet that's the same Tim Cook who has been outspoken in his defense of immigration and minority rights and in his critique of the corrosive impact of fake news on our political culture.

Yes, it's complicated. As the *Financial Times'* John Thornhill notes, today's tech billionaires "have publicly stated ambitions way beyond making money and are extending their reach into areas such as transport, health and education."[11] For all their all-too-human shortcomings, Cook, Benioff, and Bezos are exceptionally smart and, in some ways, responsible people, who, with Bill Gates and a handful of other tech tycoons, are emerging as the philanthropic Carnegies, Stanfords, Rockefellers, Vanderbilts, and Fords of the early twenty-first century. Let's also not forget Mark Zuckerberg, who, in December 2015, committed to giving away 99 percent of his wealth within his lifetime.

Like Gates or Zuckerberg, the robber barons of the nineteenth century weren't exactly paragons of virtue. Take, for example, the California railroad tycoon Leland Stanford, who also became the governor of California and a US senator. Before the Four Horsemen of the Apocalypse, there were the "Big Four"—the four nineteenth-century Northern Californian magnates behind the Central Pacific Railroad: Mark Hopkins, Charles Crocker, Collis Huntington, and Stanford himself. A master of the dark arts of political deal-making and bribery, Stanford exploited the monopolistic economies of scale of the Central Pacific Railroad to become one of America's wealthiest transportation magnates. And his business partner and fellow Big Four member, Charles Crocker, successfully starved out striking Chinese railroad laborers who were doing the dangerous work that no one else would do.

Yet Stanford also bestowed one of the largest endowments ever on a private university when he donated eighty-eight hundred acres of his own ranchlands and farms to set up a new college in Palo Alto. Tech-focused from the start, Leland Stanford Junior University opened, tuition-free, in 1891 and has nurtured generation after generation of entrepreneurs. Bill Hewlett and David Packard, the founders of Hewlett-Packard and the founders of the Silicon Valley financial and technological ecosystem, graduated in the 1930s, and since then Stanford's Office of Technology Licensing has supported such innovations as data analytics, recombinant DNA, and the Google search engine.

The Microsoft cofounder and longtime CEO Bill Gates, is, in some ways, the modern reincarnation of Stanford.

Having become the richest man in the world in the first half of his life through the use of often unscrupulous and sometimes illegal methods of crushing his rivals, he has spent the second half of his life giving away his billions to those less fortunate than himself. Today, Gates has become what David Callahan, the editor of the *Inside Philanthropy* website and the author of the 2017 book *The Givers*, calls a "super-citizen" of the "new gilded age"—a philanthropist so powerful that he is often able to shape the education, health care, and economic policies of governments around the world.

"Technology tycoons seem to be gradually replacing ageing oilmen as the heavy lifters of largesse,"[12] notes *Financial Times'* Anjana Ahuja. And Gates is the heaviest lifter of them all—a man who, like his friend Warren Buffett, is determined to give away all his money before he dies. And, as Callahan points out in *The Givers*, Bill Gates and his Gates Foundation have become the model for "the disrupters"— a new generation of young tech super-citizens including Facebook billionaires like the company's former president Sean Parker, its cofounder Dustin Moskovitz, and Mark Zuckerberg himself.[13]

So how should today's tech billionaires be giving away their money? What makes a good Silicon Valley super-citizen in our new gilded age?

It's Complicated

It's certainly complicated. Just as whole books have been written about the morally complex philanthropy of the Stanfords, Carnegies, Rockefellers, and Vanderbilts, so a book should, and no doubt will, be written about Mark Zuckerberg's much-publicized and highly controversial commitment to the public good. Yes, in 2015, the thirty-one-year-old Zuckerberg and his wife, Priscilla Chan, to great public fanfare, committed to donating a breathtaking 99 percent of their then $45 billion wealth to "charity." Later it became clear that their fortune will be put into a limited liability company that isn't even officially considered a charity by the IRS—thus allowing them to invest their money in for-profit businesses at a low tax rate. So, for example, the $600 million that Chan invested in Biohub, a scientific research organization employing some of the best researchers at Bay Area universities, would allow her to own the intellectual property it develops around vaccines and drugs for infectious diseases like HIV and Zika; and she would also, in theory, profit from its sale.

As the *Financial Times*' Edward Luce explains, Zuckerberg—as an aspiring super-citizen, has pulled off "something remarkable. He has ensured that he will pay no tax on transfer of his estimated $45bn in wealth while being showered with accolades for 'giving it all away.'"[14] Then there's the six-thousand-word mission statement Zuckerberg laid out in February 2017 in which he announced Facebook's commitment to solving many of the world's problems, including the economic crisis of the local news industry.[15] Some have interpreted this essay as a prelude to Zuckerberg's entry into politics. Perhaps, like Leland Stanford, he will one day become a California governor or a senator. Zuckerberg certainly already possesses Stanford's moral ambiguity. As Steven Waldman, the former advisor to the chairman of the

Federal Communications Commission (FCC), notes, Zuckerberg's long-winded mission statement, much of which is a thinly veiled advert for the supposedly ethical utility of Facebook, failed to take responsibility for Facebook's central role in this crisis and failed to donate the resources to solve it.

Waldman contrasts Zuckerberg with the steel magnate Andrew Carnegie. "The nineteenth century robber baron Andrew Carnegie gave away most of his wealth later in life. Carnegie built almost 3,000 libraries. All Mark Zuckerberg, Larry Page, Sergei Brin and Laurene Powell [widow of Steve Jobs] have to do is fund 3,000 journalists," Waldman observed. "If the leaders of these companies put the equivalent of just 1 percent of their profit, for five years, to the cause, local American journalism would be transformed for the next century."[16]

Waldman is right to use Carnegie—the author of the influential 1889 article "The Gospel of Wealth," a kind of More's Law pamphlet for the rich that called on them to invest their wealth in improving society—as a role model for our twenty-first-century tech barons. In business, Carnegie was anything but an angel. His steelworks were the sites of some of the most violent labor clashes of the late nineteenth century. After a long history of union-busting in their factories, his partner Henry Clay Frick even hired an armed militia in 1892 to put down a strike at the Homestead, Pennsylvania, site; the confrontation resulted in the deaths of eighteen people. And yet after his retirement Carnegie dedicated himself to giving away almost all his wealth during his lifetime, a gospel taken up today by Bill Gates and Mark Zuckerberg. This self-educated and self-made steel mogul actually helped finance more than twenty-five hundred free public libraries in his lifetime, including libraries in San Francisco and Oakland, as well as Palo Alto's first public library building, opened in 1904. When Carnegie died, in 1919, he had given away about $350 million, some 90 percent of his wealth, to charities, foundations, schools, and libraries—a sum that would be worth almost $4.8 billion today.

But rather than how much, it's *how* Carnegie gave away his money that should be so instructive for today's tech billionaires. As his "Gospel of Wealth" piece argued, it was the duty of the wealthy to improve society. "They have it in their power during their lives to busy themselves in organizing benefactions from which the masses of their fellows will derive lasting advantage, and thus dignify their own lives," he wrote about the responsibility of the rich man to help those less fortunate.

With great power, Carnegie strongly believed, comes great responsibility. His massive investment in libraries captured this sense of social responsibility. First, by financing so many free public libraries around America, he was investing in his fellow citizens so that they could, as he had done, as a young Scottish immigrant, make something of themselves by educating themselves. His libraries, he said, were designed for "the industrious and ambitious; not those who need everything done for them, but those, who, being most anxious and able to help themselves, deserve and will be benefitted." Second, Carnegie was investing in communities and in public space. His strategy was to provide the capital as long as the local authorities matched his money by providing the land and the necessary budget to staff and operate the library. One of his greatest legacies, then, was a collection of urban architectural monuments—from baroque to Spanish Renaissance to Italian Renaissance—that now grace many American downtowns.

What is missing from the philanthropy of most tech billionaires today is a similar kind of civic engagement. Yes, there is a kind of arms race of giving going on right now, with these billionaires competing to see who can give away the most money. But rather than borrowing the Carnegie model of investing money thoughtfully in the improvement of society, much of today's tech philanthropy seems to simply reflect the particular enthusiasm of the donor or the donor's spouse. Having frittered away $500 million in a wasted investment in New Jersey schools, Mark Zuckerberg, for example, seems to have settled on investing $3 billion of his wealth in the herculean effort to "cure, prevent and manage all disease."[17] Jeff Bezos, in contrast, is an avid space colonialist who even has a plan to set up an Amazon-like delivery service for "future human settlements" on the moon.[18] And Google cofounder Sergei Brin is investing between $100 million and $150 million in building the world's biggest aircraft to deliver supplies for humanitarian missions. It will, however, also be designed to serve as a luxurious intercontinental "air yacht" for Brin's friends and family.[19] Andrew Carnegie, no doubt, would be turning in his grave at such an indulgent use of valuable resources.

So how should Brin, Bezos, and Zuckerberg be investing their money in society? Perhaps they should adopt General Colin Powell's famous "you break it, you own it" view of war. That check to support the three thousand journalists, which Steven Waldman says is essential to save local American news, should probably come from Brin, whose search engine has played a central role in undermining the business model of traditional newspapers. And Brin, whose company, you'll remember, is even trying to privatize geography through its dominant map software, might replicate the work of Andrew Carnegie and invest part of his fortune in the creation of public space. But while Carnegie invested his wealth in the physical architecture of downtown libraries, Brin should figure out how to create digital public space where our privacy is guaranteed and there is no advertising. That shouldn't be too hard for him. After all, it was the original vision of Google when he and Larry Page created their search engine as Stanford graduate students in 1998.

Rather than funding interplanetary travel, Jeff Bezos should invest his money in figuring out how to employ people in an automated future, which Amazon—with its billions of dollars of investment in delivery drones and robot-operated fulfillment centers—is pioneering. This will be the greatest question of the twenty-first century and Bezos has both the financial resources and the intellectual discipline to confront this astonishingly complex issue. I sense that Bezos is only now really breaking into his stride as a public figure. Like Steve Jobs, he is a man of superhuman ambitions and abilities. With Apple, Amazon—which Jeff Bezos founded in 1994—remains the most remarkable American company of the last half century. But if Bezos wants to be remembered by future generations, he should worry about jobs rather than e-commerce. Building the "Everything Store" is one thing; fixing the future is quite another.

So that's my tweet to Bezos about where he should put his philanthropy. "Make sure we all have jobs in the future, Jeff. Easy."

Instead of throwing billions of dollars into quixotic ventures that will supposedly let us live forever, Mark Zuckerberg would be better off trying to confront the global problem of technological behavioral addiction—the "Facebook hook" and "Instagram hook" that the New York University psychologist Adam Alter argues are shrinking our attention span to less than that of a goldfish. Zuckerberg might, for example, invest some of his time and resources in Tristan Harris's nonprofit movement Time Well Spent, with its commitment to introducing a new Hippocratic Oath for software developers against the invention of addictive apps like Facebook, Snapchat, and Instagram.

In July 2017, Reid Hoffman and his friend Mark Pincus, the cofounder of the online game developer Zynga, introduced a new group to hack the Democratic Party called Win the Future (WTF). Their goal, according to the online tech website Recode, is "to force Democrats to rewire their philosophical core, from their agenda to the way they choose candidates in elections."[20] But WTF puts the cart before the horse. Rather than introducing the Silicon Valley culture of creative destruction into politics, Hoffman and Pincus should be focusing their resources on taking responsibility for the disruptive impact of technology on society. I've already suggested that Jeff Bezos should invest some of his stratospheric wealth in figuring out what jobs people will do in an automated future. And it would also be wise for Silicon Valley notables like Reid Hoffman and

Mark Pincus to work closely with conventional politicians, like the lieutenant governor of California, Gavin Newsom, who, you'll remember, is now openly warning about the "tsunami" of unemployment and inequality set off by the digital revolution.

The "you break it, you own it" model is already being pioneered by one of Silicon Valley's most responsible tech philanthropists. In 1995, a former IBM programmer named Craig Newmark founded Craigslist to enable Bay Area people to freely buy and sell local services and products on the internet. Today the site, which is valued at more than a billion dollars, covers twenty countries and services twenty billion page views per month. Yet Craigslist, by giving away free local listings to its users, unintentionally undermined the classified advertising business model of local newspapers. Newmark's response to this disastrously unintended consequence of Craigslist's success has been to invest part of his wealth in a foundation to help journalism reinvent itself in the digital age. In December 2016, for example, his Craig Newmark Foundation invested a million dollars to fund a university chair in journalism ethics.[21] And Newmark's foundation is also a contributor—with John Borthwick's Betaworks (and Zuckerberg's Facebook, to be fair)—to a $14 million fund at the City University of New York, run by the prominent journalism professor Jeff Jarvis, that is dedicated to fighting fake news by increasing trust in journalism.[22]

I'm not sure that Newmark would approve of Andrew Carnegie's rather unforgiving self-help doctrine, but he nonetheless shares Carnegie's commitment to reinvest his wealth back in society. F. Scott Fitzgerald famously wrote that there are "no second acts in American lives." But he couldn't have been more wrong about the lives of successful tech entrepreneurs. What's needed in Silicon Valley are more Craig Newmarks willing to invest the second acts of their lives in down-to-earth schemes to win the future.

The View from Nowhere

Looking to escape from Silicon Valley's out-of-control rents and endemic traffic problems, many Bay Area residents have fled over the Bay Bridge to Oakland, the industrial port that is now transforming itself not only into the Bay Area's liveliest city but also, perhaps, into its conscience.

"There's no *there* there," the early-twentieth-century American writer Gertrude Stein famously said about her hometown of Oakland, the inchoate industrial port over the Bay from Silicon Valley.

You'll remember Huw Price's philosophical quest to find Archimedes' point and its authoritative "view from nowhere"—that vantage point from where one can not only objectively observe the world, but also change it. The perspective from early-twenty-first-century Oakland might represent that view from nowhere. It's just about the ideal spot not only for getting an honest take on Silicon Valley but also for beginning to build a more socially responsible alternative ecosystem of investors, technologists, and start-ups.

When I first came to the East Bay in the early eighties to study at UC Berkeley, much of Oakland—especially the ravaged downtown area around the old Paramount Theatre on Broadway, the thirty-five-hundred-seat art deco movie house that finally closed in 1970—was like a war zone, with burned-out buildings, boarded-up stores and offices, derelict industrial factories, and crime-infested streets. Today, however, Oakland's downtown is being reinvented in the same way that Estonia and Singapore transformed themselves from sleepy backwaters into global hubs of digital innovation. Innovative new businesses, people, and ideas are reinventing this decayed industrial shell into one of the most connected cultural and economic spots in America. But in Oakland, unlike other parts of the San Francisco Bay Area, especially Silicon Valley, not everyone moving in wants to become a tech billionaire.

Today, the area around the Paramount, now renamed the Uptown neighborhood, is buzzing. The palatial old theater, which, when first built in 1931, was the largest movie house on the West Coast, has been reinvented as the nonprofit home of the Oakland East Bay Symphony and the Oakland Ballet. This Uptown area—which is also home to Andrew Carnegie's original turn-of-the-twentieth-century Oakland library, now transformed into the African American Museum and Library—is packed with hip restaurants, factory-style lofts, tech start-up zones, and all the other accoutrements of successful urban regeneration. It's certainly a long way from the sixteenth-century Hall of Cambridge University's Trinity College—with its Renaissance features that have barely changed in five hundred years—to this part of downtown Oakland, with its almost complete reinvention over the past decade.

One of the most impressive examples of this revival lies a couple of blocks north of the Paramount, on the corner of

Broadway and Twenty-Second Street, in a restored three-story building that—perhaps more than any other single building in the Bay Area—is charting a realistic map to fix the future. Between 1920 and the end of the twentieth century, this building was just another anonymous office block in a decaying part of a forgotten city. But in 2012 the vacant building was acquired by two of tech's most innovative thinkers and transformed into the beginnings of a new tech ecosystem that could turn out to be as socially valuable today as Carnegie's libraries were in the late nineteenth century.

In *The Internet Is Not the Answer*, I wrote about a four-story private club in downtown San Francisco, called the Battery, that, in spite of its boasts of inclusivity and openness, is actually a nauseatingly exclusive gentleman's-style retreat for a tech elite of mostly rich white young men. The four-story reinvented building over the Bay in downtown Oakland is everything that the Battery claims to be, but isn't. Created by Mitch Kapor—the founder of the software giant Lotus, which was acquired in 1995 by IBM for $3.5 billion—and his wife, Freada Kapor Klein, a longtime tech activist and reformer, it is part venture capital firm, part nonprofit hub for tech activists, part resource for local students and aspiring businesspeople, and part inclusive community center, restaurant, and café.

The forty-five-hundred-square-foot building, which opened in July 2016 after a two-year redesign project that totally gutted the old office block and transformed it into an airy, four-story workspace designed by its architects to blend "high-tech and humanism," is the home of the Kapor Center for Technology and Social Impact. It houses all three of Mitch and Freada Kapor's tech-for-the-social-good initiatives: Kapor Capital, the Kapor Center for Social Impact, and the Level Playing Field Institute.

None of these initiatives are new. The Center for Social Impact is a ten-year-old nonprofit organization designed to attack social and economic injustice through mostly Oakland-based initiatives to increase diverse tech start-ups, improve access to capital, and improve STEM (science, technology, engineering, and math) education. The Level Playing Field Institute is a pipeline program established by Freada Kapor Klein in 2001 and designed to funnel underprivileged local kids, through its Oakland-based Summer Math and Science Honors program (SMASH), into tech firms. And Kapor Capital is a seed-stage investor in both for-profit and nonprofit tech start-ups; between 2015 and 2017, it has already invested $40 million in education that focuses on improving access for minorities. Forty-four of these seventy-nine investments (59 percent) have a founder who is a woman or a person of color; 42 percent of their first-time investments have a female founder; and 28 percent have a founder who is a member of a racial minority. In contrast with all too many Silicon Valley start-ups that have no clear benefit to the broader community, all of Kapor Capital's investments have a social purpose. There's Pigeonly, for example, founded by the formerly incarcerated Frederick Hutson, which helps inmates make calls from prison. There's Phat Startup, a company focused on helping young inner-city people who identify with urban culture succeed in developing technology careers. Then there's the senior-care start-up, Honor, and BeneStream, a health-care tech company that helps employees navigate the bureaucratic complexities of the Affordable Care Act.

Together, these three parts of the Kapor network have been set up as an East Bay counterweight to Silicon Valley's mostly corrosive indifference to the impact of its disruption on the world around it. "You know that on the other side of that bridge, there are start-ups trying to solve the problems of the rich. We aim to make sure that Oakland becomes the place that is the center of solving the problems for the rest of us," noted Kapor Capital partner Ben Jealous, the former CEO and president of the NAACP and one of America's leading thinkers on technology, politics, and business. "We're here to make sure that as Oakland becomes more tech, tech becomes more Oakland."[23]

On the sunny December 2016 morning when I visit the new Kapor Center on Broadway in Uptown Oakland, its front door onto the street is plastered with a quote from the new California senator Kamala Harris. Oakland—and, indeed, all of the Bay Area—is still reeling from the election of Donald Trump a month earlier, so there's a need for a reminder that the world hadn't actually ended on November 8, 2016: WE MUST NOT BE OVERWHELMED OR THROW UP OUR HANDS. IT IS TIME TO ROLL UP OUR SLEEVES AND FIGHT FOR WHO WE ARE proclaimed the quote from Harris.

Inside the Kapor Center, that *fight* is raging. Kapor Capital is hosting a hackathon for tech start-ups focused on the creation of jobs for the underprivileged. The theater in the basement of the new building is full of young tech entrepreneurs and Kapor management, including Mitch

Kapor, Ben Jealous, Freada Kapor Klein, and their latest partner, the venture capitalist Ellen Pao, who in 2015 fought a high-profile gender-discrimination suit against the Silicon Valley blue-chip firm Kleiner Perkins Caufield & Byers. By the way, that's the firm that was cofounded by the late Tom Perkins, the notoriously insensitive billionaire tech investor who, in a 2014 letter to the *Wall Street Journal*, explicitly compared progressives to Nazis.

The $100,000 Kapor Center hackathon is won by a young woman from Chicago called Tiffany Smith. An MBA student at Northwestern University, Smith has a for-profit start-up called Tiltas, which is an online networking platform designed to connect the 650,000 people who come out of prison each year—whom she calls the "formerly incarcerated"—with job opportunities. Smith, who already is running a pilot program in a prison facility on the west side of Chicago, told me that she was inspired to start Tiltas by the experience of incarcerated friends who had found it impossible to get work after their release. Other entrants in the hackathon included an app designed to remove bias from hiring decisions, a platform that enables employers to create a diverse applicant pool for new jobs, a suite of tools helping start-up founders provide truly fair compensation to their employees, and a mobile platform connecting traditional businesses with independent labor to fill hourly paid shift work.

After the hackathon I meet with Freada Kapor Klein who, with her husband Mitch, has emerged as a kind of de facto spokesperson of the opposition to the gilded age excesses of Silicon Valley—the conscience, if you like, of the revolution. As an early investor in Uber, for example, she has been very vocal in her critique of Uber's controversial treatment of its female employees. In February 2017, after a female former Uber engineer wrote a widely read blog post exposing the systemic sexism and sexual harassment at the company, Kapor Klein with her husband, Mitch, even penned an "Open Letter to the Uber Board and Investors," which went viral on the internet, about what they called the "toxic" culture of the rideshare start-up.[24] And after revelations about 500 Startups CEO Dave McClure's systematic sexual harassment of female entrepreneurs, Mitch Kapor—a limited partner in one of the 500 Startups investment funds—publicly stated his intention to ask for his money back from the fund.[25] "Tech must change," the influential TechCrunch journalist Josh Constine wrote about the corrosive culture at Uber, "and more people like the Kapors must stand up for that change."[26]

Kapor Klein, who never seems to appear in public without her dog Dudley, tells me that she grew up in Los Angeles in an activist family and went to UC Berkeley in the early countercultural seventies. In 1984 she went to work at Lotus, where her job description was to create the "most progressive employer in the United States." At Lotus she cofounded an investigative group on sexual harassment, the first of its type in the United States. And she was behind Lotus's sponsorship of an AIDS walk in 1984—Lotus was the first US corporation to associate its brand with the fight against AIDS.

I ask her what happened between 1984 and 2016 in tech to make the creation of the Kapor Center so necessary.

"Many steps forward," she acknowledges, "and a couple of huge steps back."

Those "huge steps," she says, include what she calls Silicon Valley's remarkably "self-serving" faith that it's a "perfect meritocracy." Nobody in Silicon Valley, she says, recognizes his own incredible luck in happening to find himself in the wealthiest bubble in human history. "A little humility would go a long way," she says about those young, privileged men, like Sam Altman, on the other side of the Bay.

There's also a complete lack of empathy in Silicon Valley, she says. Particularly a failure of the almost exclusively white male executives to take responsibility for the "unintended consequences" of the tech boom—especially the income disparity, the homelessness, and the economic dislocation of a predominantly minority population that is transforming the San Francisco Peninsula into a nineteenth-century tableau of shockingly explicit inequality.

I ask her what distinguishes her and Mitch from the other plutocrats of Silicon Valley. They've obviously invested tens of millions of dollars in the Kapor Center. But it's more than the money. After all, Mark Zuckerberg had, in theory at least, given $45 billion of his personal wealth to "charity." And every Silicon Valley tech notable—even Travis Kalanick, the Randian former CEO of Uber—talks a good game when it comes to respecting the rights of minorities and women. But what Kapor Klein is doing in Oakland, particularly in terms of the networks the Kapor Center is building with the local community, seems so much more genuine than Zuckerberg's meretricious altruism.

"Why should I trust you?" I ask.

"We haven't lost sight of where we've come from. And we've put ourselves in different circumstances," Kapor Klein says about their commitment to this ambitious project. She explains that they've moved from their expensive home in San Francisco's exclusive Pacific Heights to a condo in Oakland's Jack London Square, only a mile away from their Broadway office.

Trust us, she is saying. And I do trust her. Certainly more fully than I trust Sergei Brin, Jeff Bezos, or Mark Zuckerberg. What Freada Kapor Klein, Mitch Klein, Ben Jealous, Ellen Pao, and their team are doing in establishing the Kapor Center, setting up Kapor Capital, and investing in socially valuable tech start-ups like Tiffany Smith's Tiltas is immensely valuable. The future, of course, can't be fixed overnight. But the Kapor Center is seeding a new kind of East Bay countercultural explosion against a new American ruling caste mostly lacking in empathy or responsibility.

The center is helping to establish the foundations of an alternative innovation ecosystem to Silicon Valley. Kapor Klein told me that her SMASH summer program has already graduated thirteen consecutive classes of low-income and disadvantaged kids who all went on to attend college. Other alternative innovation networks are also emerging in Oakland, with the goal, in Mitch Kapor's words, to "do tech differently."[27] These include Hack the Hood, a group that prepares low-income youth of color for a career in tech by hiring them to build small websites, and Black Girls Code, which is focused on improving the STEM skills of young women. In February 2017 the Kapor Center, along with the city, launched an Oakland Startup Network to support Oakland-based tech entrepreneurs. This initiative won the backing of the prestigious Kauffman Foundation, which is

working to establish Oakland as a model for diverse entre-
preneurship and provide other communities with a replicable program.

What's emerging in 2017 Oakland is the beginnings
of an ethical tech movement that will contrast vividly with
what's going on over the Bay in Silicon Valley. It's similar
to the leading role played by the neighboring East Bay city
of Berkeley in the emergence of a healthier food economy
during the seventies and eighties. Food in postwar America, you'll remember, was mostly processed, uniform, and
unhealthy. But in 1971, Alice Waters, a sixties political activist, opened a restaurant in Berkeley called Chez Panisse that
prioritized slow cooking with local, market-fresh, organic
products. Since then, Waters has been joined by other local
East Bay residents, like the food writer Michael Pollan, in
popularizing what has become known as the Berkeley Food
Movement. Today, the ideas about fresh, high-quality local
produce that Waters and Pollan developed have become so
mainstream that Amazon acquired the organic grocery chain
Whole Foods for $13 billion in 2017. Let's hope that, over
the next few years, Mitch and Frieda Kapor's ideas about
"doing tech differently" will spread with the same fecundity.

You'll remember what the Greek mathematician
Archimedes wrote about his "view from nowhere": "Give
me somewhere to stand and I shall move the earth." In its
commitment to building an alternative ecosystem to Silicon
Valley, the Kapor Center and its network of local partners
are creating that *somewhere to stand*. Of course, neither Oakland nor the Kapor Center is unique. The former AOL
CEO Steve Case, for example, is bringing the idea of social

investing to the rest of America through his Revolution Ventures fund. And all over America there are similarly innovative entrepreneurs working on a more socially responsible tech ecosystem than that of Silicon Valley. One particularly impressive young social entrepreneur Case introduced me to is Ross Baird, the founder of Village Capital in Washington, DC, a fund that is also focused on investing in tech start-ups run by minorities and women.

But Oakland, because of both its geography and its history, remains special. A century after Gertrude Stein's heartless remark about her hometown, things have changed in the city on the other side of the San Francisco Bay. Now there is a *there* there. You can find it in Uptown, a couple of blocks north of the old Paramount Theater.

Chapter Nine

WORKER AND CONSUMER CHOICE

The Strike

On a map, it isn't far from Silicon Valley to Hollywood. I am headed down to the studios of 20th Century Fox, one of Hollywood's "Big Six" movie companies. So it's just a fifty-minute flight from San Francisco to Los Angeles and then a short cab ride to the Century City location a few miles west of Beverly Hills, where the 20th Century Fox lot—originally developed by the Fox Film Corporation in 1916 from three hundred acres of Los Angeles farmland—is situated.

But geography can sometimes be deceptive. Measured in anything except miles, it's actually quite a distance from the technologists of Silicon Valley to the content makers of Hollywood. It's a relationship, observed the veteran technology writer Michael Malone in 2017, that's "on the rocks, a victim of seemingly irreconcilable differences."[1] Lingering issues such as piracy, Malone explains, have "generated mistrust" between Hollywood and Silicon Valley. The biggest problem, however, is what Malone calls the "unprecedented success" of Big Tech compared with the economic struggles of the movie studios. "There's no way Hollywood can win under the current rules," Malone observes about a new digital distribution system monopolized by such companies as Amazon, Google/YouTube, and Apple. In this bleak conclusion, Malone might have also been describing the impact of Silicon Valley on other creative industries—particularly the recorded music business, whose global revenue has been cut in half over the last twenty-five years.

As my taxi winds its way through the 20th Century Fox lot, we drive past movie sets designed as backdrops to the telling of Hollywood stories. There is even a set of what appears to be a cobbled street of nineteenth-century industrial New York City—the sort of grimy urban backdrop where one could imagine striking garment or meatpacking workers confronting thugs hired by capitalist bosses. In Hollywood, nineteenth-century neighborhoods really are full of twenty-first-century things.

I have come to Century City to meet with the movie producer, music promoter, and entrepreneur Jonathan Taplin, whom I've known since the mid-nineties, when he founded Intertainer, the internet's first video-on-demand service. We are sitting outside the Fox Commissary—the oldest staff canteen in Hollywood, which, over the last eighty years, has entertained such stars as Marilyn Monroe, Elizabeth Taylor, and Richard Burton. Originally called the Café de Paris and doubling as a restaurant set, the Commissary is an art deco building that still boasts a set of original 1935 murals featuring Shirley Temple and Will Rogers.

In this theatrical setting, Taplin has some appropriately dramatic news for me about the creative industry's response to what is widely perceived as YouTube's exploitative business model. "You have to swear not to tell anyone," he confides. "It's a total secret. Nobody knows about this yet."

"Absolutely, Jon," I reply, putting my hand on my heart jokingly. "I mean, who would I tell?"

You'll remember that the Los Angeles–based Jonathan Taplin—the serial entertainment entrepreneur who produced Martin Scorsese's breakthrough 1973 classic *Mean Streets* and managed some of Bob Dylan's early tours as well as George Harrison's *Concert for Bangladesh*—argues that the digital revolution has triggered a "massive reallocation" of $50 billion a year from the creative industry to winner-take-all internet companies such as YouTube and Facebook. Like Freada Kapor Klein, Taplin is a central figure in the growing opposition to Silicon Valley. Like the Oakland-based tech activist, he too is a veteran of the Vietnam War protests and civil rights struggles of the sixties and seventies. In contrast, however, with Kapor Klein's goal of building an alternative investment ecosystem to Silicon Valley, Taplin's focus is more directly political. His 2017 book, *Move Fast and Break Things*, borrows its title from Facebook founder Mark Zuckerberg's notorious mantra to disrupt as much as possible without taking responsibility for any of the resulting damage. Taplin reserves his sharpest critique for Silicon Valley's libertarian ideology, with its fetishization of free market forces, which has enabled this new economy to so radically disrupt old business models and economic practices. And he has become a kind of informal labor organizer, encouraging the resistance of musicians and filmmakers to these controversial new models and practices.

"It's an artists' strike," Taplin continues, lowering his voice conspiratorially. We might as well be sitting on that nineteenth-century New York City movie set in the 20th Century Fox lot and plotting the industrial action of workers against their exploitative bosses. Taplin is talking about a full-scale strike of musicians opposed to the economics of YouTube, with its controversial business practice of paying artists only around a sixth of what other services like Apple and Spotify pay. The strike would be modeled on Taylor Swift's 2014 withdrawal of almost all her music from Spotify and her 2015 decision to pull her bestselling album *1989* off Apple Music because the new streaming service wasn't intending to pay artists in the first three months of its existence.[2]

Over the last couple of hundred years, labor strikes have been one of the most effective ways for employees to force reform of their working conditions and pay. If we go back to the New York City of the late nineteenth and early twentieth centuries, for example, strikes often triggered deadly violence between workers and the authorities. In the summer of 1877, after New Yorkers came out in support of a national railway workers' strike, the authorities—fearing a proletarian revolution—called out the National Guard, who descended on the protesters with guns and clubs. In 1886, streetcar conductors, having walked out in support of a twelve-hour workday and a dinner break, responded to police beatings by setting their streetcars on fire. Most notoriously, in 1909, in the largest work stoppage in US history, ladies' garment workers from New York's Triangle

Shirtwaist Company walked out for shorter hours, better pay, and safety. In response, private detectives organized scab labor, and owners hired prostitutes to start fights with striking workers. On Seventh Street, workers packed into Cooper Union college roared in support as Clara Lemlich—a slight, dark-eyed girl who had endured six broken ribs from a beating—urged on the demonstrators. "This is not a strike," Lemlich cried; "this is an uprising!"

Taylor Swift is no Clara Lemlich, of course—either in her personal bravery or in her economic need. But in principle Swift's withdrawal of her work from Spotify and Lemlich's withdrawal of her labor from the Triangle Shirtwaist Company aren't fundamentally different. In both situations, workers have arrived at the last resort, withdrawing their labor from the market in order to reform the system. In both cases they are using one of our five tools to fix not only their future, but the future of generations of workers and artists too.

As I write this in June 2017, Taplin's full-scale artists' strike—this mass withdrawal of their creative work from YouTube, Spotify, and other streaming services—has yet to occur. But the role of globally recognized creative artists, particularly musicians, has been very influential in raising public awareness about the injustices of the YouTube–dominated entertainment economy. In June 2016, a couple of months after I met with Taplin in Century City, a thousand of the world's leading musical acts—including Abba, Coldplay, Ed Sheeran, and Lady Gaga—wrote to the EU Commission's president, Jean-Claude Juncker, complaining that YouTube is "unfairly siphoning value" away from artists and songwriters.[3] YouTube, concluded Trent Reznor, the founder of Nine Inch Nails, has been "built on the backs of free, stolen content."

That same month, 180 artists—including Taylor Swift, Paul McCartney, and Carole King—also signed a petition calling for the reform of the 1996 Digital Millennium Copyright Act (DMCA)—the US law that gives major internet services like YouTube "safe harbor" from copyright-infringement liability as long as they respond to takedown requests from rights holders. The safe harbor provision in the DMCA is at the heart of many of the disputes about the accountability of internet companies like Facebook, YouTube, Twitter, and Instagram for the posting of offensive, fake, or stolen content. Inadvertently providing cover for these companies to avoid taking legal responsibility for their illegal content, the safe harbor provision essentially allows multibillion-dollar web publishers like Facebook or YouTube to avoid being considered "publishers" under the law.

"No provider or user of an interactive computer service shall be treated as the publisher or speaker of any information provided by another information content provider," the controversial clause in the DMCA states.

What *Fast Company*'s Christopher Zara described as "the most important law in tech" has, he says, turned into "a protector of privilege" because the DMCA doesn't make websites legally accountable for illegal content posted on their sites by third parties.[4] As a magnified version of the internet itself, the DMCA represents a tragic law of unintended consequences, explains internet historian John Naughton. Just as the prohibition of alcohol in the United

States from 1920 to 1933 triggered a massive increase in organized crime, so the DMCA has led, Naughton says, to an "astonishing rise in hate speech, harassment, bullying, revenge porn, fake news and other abuses of digital technology."[5] Yes, move fast and break things, the DMCA is saying. Yes, profit from all the content posted on your network. But no, don't worry about the consequences of this content, because you aren't accountable for any of it under the law.

Principal among those "other abuses" Naughton describes are the posting and exchange of illegal, mostly pirated content. And so opposition to the DMCA has become a rallying cry for artists who, like Taplin, see the internet as enabling that "massive reallocation" of wealth from the creative community to Silicon Valley. The June 2016 artists' petition, issued in Washington, DC, publications like *Politico*, *The Hill*, and *Roll Call*, argued that the DMCA "has allowed major tech companies to grow and generate huge profits by creating ease of use for consumers to carry almost every recorded song in history in their pocket via a smartphone, while songwriters' and artists' earnings continue to diminish."[6]

Many other artists, including Katy Perry, Mötley Crüe cofounder Nikki Sixx, and Debbie Harry, the lead singer in the seventies new wave band Blondie, have been outspoken in their calls to reform the DMCA. The problem, Debbie Harry argues, is that the DMCA has a "loophole"—the "impossible to enforce" clause in the law that enables YouTube to run huge amounts of illegal content and "not to pay artists properly." While Blondie videos on YouTube garner hundreds of millions of views, Harry argues, "none of us in Blondie will receive a fair amount of royalties from these millions of plays."[7]

This kind of celebrity resistance to both the DMCA and YouTube might not have quite the same dramatic resonance as the Progressive-era New York City street battles between striking garment workers and thugs hired by the factory owners. But it does represent a new kind of industrial action for the networked age—a mix of direct political action, coalition building, pressure on legislative and regulatory authorities, and the powerful threat of a collective boycott if nothing else works.

The boycott card is also being played by other coalitions opposed to Google's and YouTube's commercial dominance. The advertising industry is becoming increasingly exasperated with Google's and YouTube's automated plastering of online ads next to offensive content such as videos promoting anti-Semitism or terrorist groups like ISIS and al-Shabaab. This represents another example of these hugely powerful and wealthy companies refusing to be accountable for the content posted on their networks. The really troubling thing about this advertising problem is that Google and YouTube are, essentially, making money from overtly racist and violent videos and other disgustingly inappropriate content.

In March 2017, a number of blue-chip firms—including the UK offices of McDonald's, L'Oréal, Volkswagen, Audi, Vodafone, Sky, HSBC, Lloyds, and Royal Bank of Scotland—all suspended their advertising on Google and YouTube because their ads appeared next to videos featuring hate preachers and the former Ku Klux Klan leader David Duke.[8] The French advertising company Havas, the world's sixth-largest marketing company, even went as far as to pull all its UK clients' advertising from Google in an attempt to get the Silicon Valley company to police the content on its platform more carefully. And a few days after these European boycotts made news, a number of major American advertisers, including Starbucks, AT&T, Walmart, Verizon, and Johnson & Johnson, joined the defection, announcing an end to their ad spending on Google until the system is made more accountable.[9]

These actions have the implicit support of Sir Martin Sorrell, the CEO of WPP, the world's largest marketing company, who remarked that Google and Facebook have "the same responsibilities as any media companies" and could not "masquerade" as mere technology platforms.[10] Sorrell, with whom I've spoken at great length over the years about this problem of responsibility and accountability, is absolutely right. The fundamental problem about not just Google and YouTube, but also Facebook, Instagram, Snapchat, and many other Silicon Valley companies, is their unwillingness to grow up and assume the complex responsibilities of media companies. This means not only ensuring that the content on their networks isn't stolen or hateful, but also guaranteeing that advertisers don't find their messages attached to offensive or illegal content that tarnishes their brands.

The work of all these industry coalitions—from angry music artists to untrusting advertisers and marketing companies—is critically important. Nothing will be reformed unless both political and commercial pressure is placed on the private superpowers to reform their behavior and business models. If enough pressure is exerted, things will change; if not, they won't. So, less than twenty-four hours after Taylor Swift's 2015 threat to boycott Apple Music, the company changed its policy and promised to pay artists for streaming, even during a customer's free trial period. The announcement even came with an apologetic personal tweet from Eddy Cue, the head of Apple Music.

We hear you, @taylorswift13 and indie artists. Love, Apple, Cue wrote on Twitter the day after Swift threatened to boycott his service.

And immediately after the March 2017 advertising boycott of YouTube, Google's chief business officer announced "expanded safeguards" to protect advertisers, including "hiring significant numbers of people" to enforce the notice-and-takedown process, as well as using AI to speed up the process of tagging offensive content.[11] The boycott also resulted in Google's introducing what it called third-party "brand safety" features for its customers, enabling them to track their advertisements to see where they appear on YouTube.[12] Then in June 2017, Google's general counsel announced four more steps YouTube was taking to fight online terror, including more aggressively identifying and taking down flagrantly offensive videos, giving advertisers more control on where their ads appear, and hiding questionable content so that it can't be monetized or recommended.[13]

While some advertisers remained skeptical about these policy changes, what the boycott shows is that, if enough pressure is applied, Google and YouTube will, at least, try to address the issue and make themselves more accountable. Yes, as Rob Norman—the chief digital officer of the WPP-owned Group M, one of the world's largest buyers of online media—acknowledges, the filtering of inappropriate content is a "complex problem" equivalent to "threading the needle." But Norman is hopeful that Google can fix this problem. "They are one of the world's outstanding science companies," he says, "and if anyone can do this, they can."[14]

Given that six out of ten people now get their news from social media, Facebook has also been heavily criticized for its unwillingness to take responsibility for the content posted on its network. The distribution of fake news has been a particular concern, with Mark Zuckerberg even being forced by many critics to deny that Facebook was to blame for the election of Donald Trump. "No, we are a tech company, not a media company," Zuckerberg said immediately after the November 2016 American presidential election.[15] And yet, for all Zuckerberg's denials of a self-evident truth, Facebook is—as I noted earlier—increasingly behaving like a quintessential media company in its aggressive attempts to fix the problem of fake news. In December 2016 it announced a partnership with independent fact-checkers to filter fake news from its feeds. In March 2017 it launched a third-party fact-checking tool that will alert users to "disputed content"[16]—a feature that Google also introduced a couple of weeks later.[17] Then, in April 2017, after criticism from German chancellor Angela Markel about the dangerous political ramifications of fake news, Facebook—which has twenty-nine million German users—even took out full-page ads in Germany's biggest newspaper explaining how to combat this content.[18]

Boycotts, political pressure, public criticism, and even the threat of a strike do work, then, in forcing Silicon Valley companies to try to fix the most egregious consequences of their services. As Group M's Rob Norman notes, these aren't simple problems, and they don't have easy answers. But if anyone can solve the problems of fake news and the attachment of offensive content to advertising, it's Facebook and Google. Yet the only way this will happen is by challenging their bottom line. When threatened with a loss of revenue, these multibillion-dollar businesses will act and make themselves more accountable. Otherwise, they won't.

The challenge, *our* challenge, then, is to maintain the public conversation about these issues so that everyone— from advertisers to governments to consumers—is aware of how unaccountable the system is becoming. But this sometimes requires great tenacity, particularly when that conversation is only just beginning. It demands somebody crazy enough to single-handedly take on a multibillion-dollar industry.

Sex, Drugs, and More's Law

Offstage, rock 'n' roll artists are usually anything but rebels resisting powers larger than themselves. But David Lowery— the guitarist, songwriter, and vocalist who founded the nineties rock bands Camper Van Beethoven and Cracker—is unlike most rock 'n' roll artists. Rather than just singing about revolution, he is actually perched on the ramparts and leading the rebellion against the new ancien régime.

Lowery is a genuine rebel—a guy just crazy enough to personally take on a million-billion-dollar industry. As the *New York Times* explains, he "has come to represent the anger of musicians in the digital age."[19] It's an anger stoked by Jonathan Taplin's "massive reallocation" of wealth, which has resulted not only in the creation of unimaginable fortunes in Silicon Valley but also in the worse than halving of global recorded music sales over the last fifteen years, from $14.6 billion in 1999 to $7.02 billion in 2015.

Yes, business models and industries change, and nothing lasts forever, particularly in today's perpetual Schumpeterian storm of digital disruption. And yes, the success of streaming services like Spotify, Apple Music, YouTube, and Rhapsody—which now make up 34.5 percent of overall sales—seems, over the last couple of years, to have finally halted the death spiral of the recorded music economy. Between 2015 and 2016, for example, overall music streaming increased in the United States by 68 percent, with revenues on subscription services rising 114 percent to $2.5 billion.[20] But this new business model of streaming all-you-can-eat content in exchange for a monthly subscription fee, while great for consumers and the labels, has turned out to be anything but an ideal solution for creative artists.

The problem—as Debbie Harry, Katy Perry, and Lady Gaga all remind us—is that many of the streaming services, particularly YouTube and Pandora, pay such miserly royalty rates. So, as the *New York Times* notes about the streaming market, "the music industry finds itself fighting over pennies while waving goodbye to the dollars." Things have become so skewed that the niche market in high-margin vinyl records is creating more revenue for the music industry than all those billions of views on YouTube. "In [the] shift to streaming," the *Times* concludes, the "music business has lost billions."[21] As David Lowery put it in a typically acerbic blog post, "My Song Got Played on Pandora 1 Million Times and All I Got Was $16.89, Less Than What I Make from a Single T-Shirt Sale!"[22]

And so Lowery, who also teaches music economics at the University of Georgia, has become the public face of the angry musician. Having launched his second career as an outspoken critic of large internet companies—in what *Billboard* describes as his "now famous" "Meet the New Boss, Worse Than the Old Boss" speech at the 2012 San Francisco Music Tech Summit[23]—Lowery has fought an unrelenting one-man war against everything from the exploitative business models of Pandora and YouTube to the immorality of online music theft. He is, according to *Bloomberg*, "one cranky rocker" who is taking on "the entire streaming music business."[24]

Like Taplin's, David Lowery's goal is to raise consciousness about the injustice of a new economic system in which musicians and songwriters are being so callously exploited. But Lowery is more than just a gadfly buzzing angrily around the new digital establishment. In late 2015 he launched a $150 million class action copyright-infringement lawsuit on behalf of other songwriters against Spotify, the $8 billion Swedish-based streaming service. The suit, based on Lowery's own painstaking research, asserted that Spotify hadn't acquired the mechanical licenses for much of the musical content it was streaming. Spotify, in Lowery's words, was a "copyright infringement machine" that was essentially stealing the work of songwriters like himself.[25] A couple of months later he filed a similar suit against Rhapsody, requesting that it too pay all songwriters and artists for the use of their work. As with the advertising industry's boycott of Google, Lowery was making these streaming companies responsible for their actions. He was forcing accountability on an industry that was breaking all the rules.

While these suits remain ongoing, Lowery's ferocious bellicosity has inspired young artists to believe that they too can change the system. According to the prominent music lawyer Chris Castle, Lowery has "changed the conversation" by showing that the only way—at least in America—to alter the system is through the courts. "Legislation," the Austin, Texas–based Castle told me, "is no longer the solution." The only people who can impose copyrights are the creators themselves. There is no guardian angel, no Margrethe Vestager in Washington, DC, Castle suggests, who will do it for them.

Dennis Kooker, the president of global digital business at Sony Music, agrees with Castle. "Lowery has made it safe for artists to jump in," Kooker told me when I visited him at his office in midtown Manhattan. Artists particularly need to speak out about the "broken" DMCA safe harbor law, Kooker insists, although he expects the major legislative push on this to come from Europe. In spite of the paid subscription growth through Apple Music and Spotify, it's been a "brutal decade" for the music industry, Kooker acknowledges, particularly for artists. And the only way that's going to change, he says, is through the political commitment of artists like David Lowery.

After leaving Kooker at Sony's glitzy office, I take the train to Philadelphia to meet Lowery. I'm headed for a small live music venue a few blocks from the downtown Philadelphia railway station. Its dirty windows are plastered with homemade agitprop signs suggesting an ongoing war with digital music. RHYTHMS RATHER THAN ALGORITHMS, COMMUNITY NOT COMMODITY, CURATED NOT ENCODED, these signs declare.

In Philly, I'm catching Lowery on tour—between gigs in Pittsburgh and Cleveland. I buy him dinner at a noisy ribs restaurant next to the venue. He is doing two sets that evening for a private corporate party. He tells me that he normally does a hundred live shows a year and makes around $100,000 annually from his music—$50,000 from the road and $50,000 from royalties, mostly presumably from his one and only big hit, the 1993 Cracker song "Low," which sold a million copies. So much, then, for the glamour and the wealth of the rock 'n' roll life.

"The business is getting out on the road, playing shows in as many markets as you can, getting bloggers to write about you," he tells me. "What's been lost is the middle class."

What is so extraordinary about Lowery is the ordinariness of his American manners and appearance. He is reassuringly reasonable in his unreasonableness. If Hollywood ever made a movie about his exploits and Jimmy Stewart were still around, he'd be perfect to play the role of this *aw-shucks* everyman rebel. I ask him about his anger. What drives him to single-handedly take on multibillion-dollar companies like Spotify and Pandora? Why invest so much time in trying to save the future of recorded music?

"My grandma said, *Red hair marks the crazy ones,*" the now mostly gray-haired Lowery confesses. "It must be the Irish or Scots in me."

Crazy, perhaps, but also smart. In 2006 Lowery—a math quant as well as a minor rock 'n' roll star—was at a party in Chicago, where he met a guy called Brad Keywell, who was carrying around Nassim Nicholas Taleb's *The Black Swan*, the bestselling book about the highly improbable. They got into conversation, and with a *Black Swan*–like improbability, Lowery ended up consulting for Keywell's start-up, a special-deals e-commerce website that later become known as Groupon. Paid in stock for his work, Lowery made a million dollars after the Groupon public offering in 2011, which at the time was the biggest internet IPO since Google's in 2004.

"When it went public, it was worth $21 billion," Lowery says, putting down a half-eaten sparerib and raising an arm in the air to indicate the absurdly stratospheric valuation of the 2011 Groupon IPO. "That was totally fucking insane."

In Silicon Valley parlance, "fuck-you money" means having so many hundreds of millions of dollars—à la Peter Thiel—that you can do or say anything you like, even publicly support Donald Trump. But all David Lowery needed was $100,000 to buy him his freedom. You see, out of that million bucks he made from the Groupon IPO, he set aside 10 percent, as a kind of personal tithe, to do good. And it's this money that has not only enabled him to fight Spotify and

Rhapsody in the courts but also to reinvent himself as the cranky rocker able to take on the entire streaming business.

"The building is on fire," Lowery says, smiling sadly, referring to the shift to a streaming business model in which consumers get access to all the music ever recorded, for under $10 a month. "My job is to turn the tables."

But—whether it's building coalitions of artists or fighting tech companies in court or just raising hell on the internet—there is no magic bullet, no single solution to fix the future of the music business, Lowery acknowledges. On the one hand, he says, the current subscription services aren't sufficiently rewarding artists; on the other, he remains hopeful that YouTube will eventually shift much of its content from free access to a subscription model. So subscription is both the problem and the answer. It all depends on who gets paid what.

There is one thing, however, about which Lowery is certain. "The existential crisis of the system," he insists, "is coming." Ironically, it will be spurred on, he hopes, by ad blocking. Once everyone on a "mass scale" starts blocking ads with disruptive services like Tim Schumacher's Adblock Plus, then all the companies with advertising-based business models, particularly YouTube, will be forced into offering subscription-based services to their users. So, in the end, the market and consumer demand will drive a shift back to a paid model. And then the challenge will be to ensure that artists are paid more than the price of a T-shirt each time a million people listen to one of their songs.

If Lowery is right about the eventual demise of advertising-based models, then we are going to need innovative new internet businesses like Peter Sunde's micro-donation network Flattr, which enables more personalized commercial relationships between artists and consumers. Flattr—like the American creative platform Patreon and the Dutch micropayments news network Blendle, the so-called iTunes for news—goes beyond the centralized business model of networks like Spotify and Netflix by enabling artists and journalists to create direct commercial relationships with their consumers. These are all examples of Brad Burnham's "decentralized marketplaces" in what he and Tim Berners-Lee call the "re-decentralization" of the web.

"You get the economy you pay for," Lowery tells me. What he is noticing, however, is that a younger generation of music lovers is recognizing the importance of paying for content. "My students get it," he tells me, smiling for the first time. "They understand that nothing is free."

The *New York Times* technology columnist Farhad Manjoo shares Lowery's tentative optimism about the beginnings of a rebirth of paid online content. "Things are turning around; for people of the future, our time may be remembered as a period not of death, but of rejuvenation and rebirth," Manjoo argues, speaking of a decentralized subscription platform like Patreon that, in 2016, enabled thirty-five artists to make more than $150,000 apiece.

The internet, Manjoo suggests, might actually be "saving" rather than "killing" culture.[26] "I don't have to go out on the road and play in bars," one a cappella singer making $20,000 a month on Patreon tells Manjoo. "I can be a father and I can be a husband. This normalizes my career.

It normalizes the career of being an artist, which has never been normalized."[27]

If Manjoo is correct about this cultural renaissance, some of the credit for it should go to David Lowery, who, perhaps more than any other working musician, has directly taken on Big Tech for the right of artists to be fairly compensated for their work. Lowery is forcing accountability on the nascent streaming economy. He is making it grow up.

The Lawyer Strikes Back

The morning after David Lowery's gig in Philadelphia, I take the train up to Boston. From my hotel near the station I order an Uber to transport me to the offices of Lichten and Liss-Riordan, a law firm based in Boylston Street, the major commercial thoroughfare running through the New England city. The car comes almost instantly, its electronic image zipping along the Boston streets on my iPhone app to alert me to its arrival. For frequent travelers, ridesharing networks such as Uber and Lyft offer an incredibly convenient and cheap way to get around cities at home and abroad. Like "free" YouTube videos or the absurdly cheap Spotify subscription service, they seem almost too good to be true. Which is exactly what many of these new digital products actually are, at least from the perspective of the workers—the creative artists or drivers whose labor provides their core value.

What it means to be a worker has changed dramatically over the last century. In the industrial age, there was the "proletariat"—the armies of permanent wage laborers who worked in the factories. Today, however, in a network age of increasing inequality between rich and poor, there is the "precariat"—a growing class of impermanently employed folks who scrape out a living renting out their spare rooms via Airbnb, delivering groceries on Instacart, or driving their cars for one of the new ridesharing companies. The operating system for labor in a networked economy, where 40 percent of all American workers are predicted to become members of the precariat by 2020,[28] is changing with dizzying rapidity. But what, unfortunately, isn't changing with anything like the same speed is the way in which the law is applied to protect these badly paid, precariously employed workers from rapacious private superpowers like Uber.

"These companies," warned an April 2017 *New York Times* editorial about gig economy start-ups, "have discovered that they can harness advances in software and behavioral sciences to old-fashioned worker exploitation . . . because employees lack the basic protection of American law."[29]

My Uber driver, a polite young man from Pakistan in a spotlessly clean white Toyota Prius, tells me that he's working part-time to help pay his tuition costs as an engineering graduate student at Boston University. I ask him whether he's happy working for the $70 billion ridesharing start-up founded by Travis Kalanick, who, in February 2017, was filmed swearing angrily at a destitute Uber driver for not being sufficiently self-reliant. The driver's answer to my question is ambivalent. Yes, he confirms, he enjoys the freedom of choosing his own hours; but the income, he says, isn't as much he'd hoped it would be—particularly after he's subtracted the insurance, gas, depreciation, and the other costs of running his Prius. In fact, he admits, on slow days he suspects that he ends up with less than the eleven-dollar-an-hour Massachusetts minimum wage.

"So Uber doesn't provide any benefits at all?" I ask.

"No, nothing. I'm an independent contractor," he tells me a little sadly. "I work for myself."

A pioneer of the so-called sharing or gig economy, Uber—borrowing from the Silicon Valley libertarian fantasy of absolute personal freedom—describes itself as a company that empowers people to work when and where they want, without having the restrictive commitments of a regular full-time job. Although this may be true in some ways, it doesn't, in practice, feel very *empowering* to most of the now more than 1.5 million Uber drivers in seventy countries around the world. The problem is that the gig economy's absence of commitment works both ways. Yes, the drivers don't have to commit all their time to Uber, but then Uber isn't committing anything to the drivers either. This is emblematic of an increasingly unequal economy where, in truth, multibillion-dollar start-ups like Uber and Lyft *share* none of their stratospheric value with their workers. Indeed, a July 2017 report about the gig economy by the British MP Frank Field alleges that some self-employed UK drivers are earning only £2.50 an hour working for sharing economy companies like Parcelforce and webuyanycar.com.[30]

"In reality, there is no utopia at companies like Uber, Lyft, Instacart and Handy, whose workers are often manipulated into working long hours for low wages while continually chasing the next ride or task," that 2017 *New York Times* editorial argues.[31] Or, as the *New Yorker* staff writer Jia

Tolentino bluntly put it, commenting on the cult of work celebrated by on-demand companies such as the freelance marketplace Fiverr, "The gig economy celebrates working yourself to death."[32]

My own experience as a heavy Uber user chimes with these conclusions. On my travels around the world to research this book, I've discussed the value of Uber with many of my drivers. Very few had anything good to say about the Silicon Valley start-up, and most admitted that they were looking for alternative ways to make extra money. There was, for example, the Singaporean retiree, a former marketing executive at Singapore Airlines, who had been pushed into driving an Uber vehicle because his son wanted him out of the house. He worked sixty hours a week and made 500 Singapore dollars (around $350)—"less than a cleaner," he told me disgustedly about his less-than-six-dollars-an-hour rate.

All this is anecdotal, of course—an old-fashioned method of research conducted from the back of a very new kind of taxi—but these conclusions are also confirmed by the data. One Uber drivers' group in New York City estimates that a fifth of its members earn under $30,000 *before* expenses like gas, maintenance, and insurance. And a 2017 report undertaken by the online technology resource *The Information* found that only 4 percent of drivers remain on the Uber platform a year after they begin working for the rideshare company.[33]

The problem, as I've suggested, is the law. Or, more precisely, the failure of the preexisting law to be applied to the new gig economy. And that's exactly why I'm visiting the law offices of Lichten and Liss-Riordan in Boston.

When we arrive on Boylston Street, I thank my driver and award him a five-star rating on my iPhone. I wish I could have tipped him too. But the app doesn't allow me to do that—presumably because the tip would go straight to the driver rather than to Uber.

There were, of course, no gig economy companies in Thomas More's *Utopia*. Nor were there any lawyers. The self-depreciatingly witty More, who himself practiced law in sixteenth-century London, banned them from his island. "In Utopia, everyone is a legal expert," he tells us about an imaginary society in which law had been radically democratized, "for the laws are very few . . . and they consider the most obvious interpretation of any law to be the fairest."[34] In the real world, however, especially in the United States, there is no shortage of either professional lawyers or complicated laws for them to interpret. Still, given the country's increasingly dysfunctional political system, this might not be such a bad thing. Historically, crusading lawyers like Ralph Nader have played an important role in reforming American capitalism. As the music attorney Chris Castle told me, citing the example of David Lowery's class action lawsuits against Spotify and Rhapsody, the law may today be our most effective avenue for making entertainment companies accountable. You'll also recall the central role of the Silicon Valley attorney Gary Reback in the US government's antitrust trial against Microsoft in the 1990s—a case that eventually enabled the innovation-rich Web 2.0 revolution of Google and Facebook.

I've come to Boston to meet a lawyer dubbed by one plaintiff "Sledgehammer Shannon," who is, according to

Mother Jones, "Uber's worst nightmare."[35] Shannon Liss-Riordan is the Harvard-trained lawyer who, over the last five years, has led the legal fight against the gig economy's treatment of its workers. *Politico* included her in its 2016 list of America's fifty most influential people, and *San Francisco* magazine described her as "the most reviled woman in Silicon Valley," who has "achieved a kind of celebrity unseen in the legal world since Ralph Nader sued General Motors."[36]

In person, the diminutive Liss-Riordan resembles anything but a sledgehammer. As the labor lawyer leads me into her office—which is peppered with political memorabilia from her fellow gig economy critic and political ally Massachusetts senator Elizabeth Warren[37]—she acknowledges that the digital revolution has made people work more efficiently. "I love what technology can do for us," she tells me, sounding a lot like Margrethe Vestager. "But it shouldn't be abused."

The comparison of Shannon Liss-Riordan with Ralph Nader is also useful. You'll remember that it was Nader's bestselling 1965 book exposing the fatal flaws in the Chevrolet Corvair, *Unsafe at Any Speed: The Designed-In Dangers of the American Automobile*, that ultimately undermined the global domination of the US car industry. Fifty years later, Liss-Riordan has uncovered the designed-in danger of America's latest automotive innovation—the online drive-sharing industry. But, whereas Nader identified the flawed mechanics of American cars, Liss-Riordan's crusade has focused on the flawed employment architecture of the American gig economy.

In 2013 Liss-Riordan filed a class action suit on behalf of California Uber drivers. It claimed that Uber has misclassified its drivers as independent contractors, when in fact they are official employees and thus have the legal right to benefits like workers' compensation, unemployment, and social security. The suit also claimed that Uber fixed the price of rides by promising that tips were included and then failed to pass on the extra money to the drivers. In 2015 Liss-Riordan compromised with Uber, negotiating a settlement of $84 million on behalf of 325,000 drivers in California and 60,000 in Massachusetts, which allowed the Silicon Valley company to continue employing its drivers as independent contractors. She has also pursued similar class action cases against Lyft, the food-delivery companies DoorDash and Grubhub, the grocery app Instacart, and the shipping app Shyp. In July 2015 the home-cleaning gig start-up Homejoy, a high-profile company that *Forbes* described as a Silicon Valley "darling,"[38] shut down, its CEO claiming that a Liss-Riordan worker classification lawsuit was "the deciding factor" in making the choice to close the company.[39]

Like David Lowery, all Liss-Riordan wants is accountability from these new companies. "I'm a surrogate regulator," she says, explaining that her job is really just helping to enforce the law. "The old rules still apply. The nature of work hasn't changed. People still need legal protection from employers."

Liss-Riordan's primary achievement is forcing the gig economy to grow up. Her lawsuits against Instacart and Shyp, for example, resulted in these two companies changing their employment policies and converting their contractors to full-time employees. Citing a 2015 National Employment Law Project (NELP) report, she tells me that there are now more and more on-demand economy companies that respect workers' rights, including the food-delivery services Sprig and Munchery, the personal assistant company Hello Alfred, the valet parking service Luxe, and the cleaning service MyClean.

It's not only Liss-Riordan who is fixing this problem. Many other entrepreneurs, regulators, consumers, educators, and workers are also working to build an on-demand economy that is simultaneously innovative and fair. You'll remember that four out of ten Americans are predicted to become part of the precariat by 2020—a prediction covering every industry from entertainment and media to transportation, education, legal, and health care.[40] So this is one of the great issues of our time. If we get it right, we can guarantee a decent quality of work for future generations.

It's easy, of course, to vilify a pantomime figure like the former Uber CEO Travis Kalanick, the libertarian multi-billionaire who used to sport an image of Ayn Rand on his Twitter profile. Fortunately, however, there are some entrepreneurs with a much more adult sense of responsibility and accountability. Take, for example, Glenn Kelman, the CEO of the Seattle-based internet company Redfin, which provides a web-based real estate database, employs more than a thousand agents, and conducted a highly successful IPO in July 2017. When Kelman founded Redfin in 2006, having sold his previous internet start-up for $200 million, he insisted that the new start-up employ real estate agents as full-time employees, with health benefits and 401(k) retirement contributions, rather than independent contractors. Glenn Kelman's rationale for this—in addition to sharing the wealth created by Redfin with everyone who worked for the company—was to provide better customer service. As a pioneer of this alternative model for start-ups, Kelman has even become what the *New York Times* calls an "informal counselor" for other start-up entrepreneurs also interested in shifting away from the independent contractor employment model because the shift makes both ethical and moral sense.[41] As the subtitle of that NELP on-demand economy report said, "Why Treating Workers as Employees Is Good for Business."

Some economists believe that we need to move away from the binary option of either the full-time employment model or the independent contractor model. Alan Krueger, the former chairman of President Barack Obama's Council of Economic Advisers, who now teaches economics at Princeton, believes that we are at a point today similar to that in the late nineteenth century with the emergence of an industrial workers' compensation system. But today, Krueger argued in a 2015 white paper,[42] we need a new category of worker, neither completely freelance nor full-time. The Princeton economist defines this new hybrid category as an "independent worker," arguing that gig economy workers at Uber or Lyft should qualify for some of the employment benefits of the old system—"except for the ones that don't make sense."[43]

Although Krueger's proposal sounds eminently reasonable, it is unlikely to be implemented in an increasingly hostile political environment that pits government regulators and the courts against tech disrupters like Uber or Airbnb. As with antitrust, we ultimately need regulatory protection from some of the most egregious consequences of the on-demand market. And governments all over the world are working to shape a peer-to-peer economy that will protect both workers and consumers. In 2015, for example, the Seattle City Council voted unanimously to approve a bill that allows drivers for ride-hailing apps to form unions.[44] In 2016 a London employment tribunal ruled that these drivers are indeed entitled to workers' rights, including the national minimum wage and paid holidays.[45] In 2016 Uber paid $25 million to settle a lawsuit filed by the cities of Los Angeles and San Francisco claiming that the rideshare company misled customers with a "false sense of security" about its checks on drivers.[46] In Austin, Texas, voters rejected Uber's and Lyft's proposal to self-regulate their services. *Trust us*, this 2016 local referendum asked. *No*, the voters replied.[47]

Politicians around the world are also working to responsibly regulate other areas of the sharing economy. Shannon Liss-Riordan's political ally, Massachusetts senator Elizabeth Warren, has taken on Airbnb, claiming that the $31 billion–valued home-sharing start-up is forcing up rents in large cities. In October 2016 Warren established a coalition of lawmakers from more than a dozen cities, urging the Federal Trade Commission (FTC) to "help cities protect consumers" and to study how the short-term rental market is affecting the overall rental market.[48] In November 2016 regulators in New York City and San Francisco successfully got Airbnb to establish a "one host, one home" rule for new hosts as a defense against rising rents. In an attempt to protect affordable housing, both Berlin and Barcelona have clamped down on Airbnb, with Berlin banning the renting of apartments to tourists and Barcelona aggressively cracking down on illegal rentals.[49] Even Iceland, in order to control the prices of the local market, has passed a law restricting the number of days that properties can be rented out on Airbnb.[50]

The precariat itself is also taking to the streets in order to change the system. In August 2016, drivers of UberEATS, Uber's food-delivery service, picketed London restaurants to demand that Uber pay them the London living hourly wage of a guaranteed £9.40 ($12.10).[51] And in November 2016 there was a national protest in the United States by Uber drivers demanding a $15 minimum wage.[52] In May 2016, meanwhile, the thirty-five thousand Uber drivers in New York agreed to form an organization called the Independent Drivers Guild, which would be affiliated with more traditional industrial labor unions.[53] Indeed, one of the first actions of this guild was an April 2017 petition, signed by eleven thousand drivers, requiring Uber to include a tipping option in its app.[54] And so I recently became able to electronically tip good Uber drivers like that polite young man from Pakistan who transported me to Shannon Liss-Riordan's office in downtown Boston.

Consumers are also using their collective muscle to force Uber to become more responsible and accountable. In early 2017, after the then Uber CEO Travis Kalanick was appointed to Donald Trump's economic advisory council, a #DeleteUber protest organized on Twitter resulted in more than two hundred thousand Uber customers (around 0.5 percent of its forty million users) closing their accounts.

"Try Lyft. Use a taxi, a bus or a train," advised the *New York Times*' Farhad Manjoo about kicking the Uber habit in the face of the company's never-ending toxicity. "Heck, hire a limo and a chauffeur with a golden top hat."[55]

Even Uber staffers joined in the protest, arguing in an internally circulated "Letter to Travis" that his support for Trump was coloring the ridesharing company as anti-immigrant. The protest was so effective that in January 2017 Kalanick bowed to popular pressure and quit Trump's advisory council.[56]

While all this labor activity is very encouraging, there is one problem—a flaw as potentially fatal as the mechanical defect that made the Corvair motorcar such a death trap. It all assumes that human labor—drivers, valet parkers, cleaners, and grocery shoppers—remains at the core of the twenty-first-century peer-to-peer economy. But this may be wrong. Imagine if robots replaced all this labor. Imagine if human labor itself is about to be made redundant by one of the most disruptive technological revolutions in history.

This isn't some science-fictional nightmare that Hollywood might flip into a dystopian movie. In January 2015 Uber "gutted" the robotics lab of Carnegie Mellon University to "poach" the fifty people who were working on self-driving cars.[57] As I write this, a little over two years later, Uber is already conducting trials of self-driving vehicles in both Pennsylvania and Arizona. Google, Apple, and many other conventional car companies are conducting similar trials. By the time you read this, we will be ever closer to the reality of self-driving cars on our roads.

The logic for this enormous investment in the technology of self-driving cars—particularly from Uber's point of view—is chillingly obvious. "Uber's future depends greatly on solving self-driving," the technology website *Recode* explains. "Taking drivers out of the equation would also increase the company's profits: Self-driving cars give Uber 100 percent of the fare, the company would no longer have to subsidize driver pay and the cars can run nearly 24 hours a day."[58]

And, as you'll remember, the Columbia University economist Jeffrey Sachs warns us that there is now an urgency to the issue of technological unemployment, not just in transportation, but throughout every sector of the economy. So what should we do when private superpowers like Uber make 100 percent of their profit by replacing their 1.5 million drivers with smart machines? How can we fix a future in which algorithms replace not only vast swatches of the manual labor force, but also skilled workers like lawyers, doctors, and engineers?

As the most serious long-term problem on the horizon, this is fast becoming the great debate of the twenty-first century. But it isn't an entirely new issue and may not require a new solution. Indeed, five hundred years ago, Thomas More—responding to disruptive changes in the sixteenth-century agricultural economy that created "greedy" and "fierce" sheep, which, as he so memorably described, "devour human beings"[59]—had already come up with a fix to this problem. So, to begin our conversation about this issue, let's return to the island idyll founded by King Utopus—that familiar no-place in the middle of nowhere.

Chapter Ten

Education

The Happiness of Life

In Thomas More's Utopia, there are no "languid idlers," no "cesspool of worthless swashbucklers," no "great lazy gangs of priests," no "lusty beggars."[1] Everybody, even women, works; but nobody works more than six hours a day—three hours in the morning, three in the afternoon. And the rest of the day, More explains, is left to "each person's individual discretion." There are rules, such as no "roistering" or "sloth," so people mostly spend their spare time either in "intellectual activity," such as attending educational public lectures, or pursuing their own "trade." In the evening, after a communal supper, people "devote an hour to recreation," such as gardening or exercising. Finally, before bed, Utopians either play music and games or engage in conversation.[2]

In an ideal society, More is saying, work matters—but leisure matters even more. "The chief aim of their constitution is that, as far as public needs permit, all citizens should be free to withdraw as much time as possible from the service

of the body and devote themselves to the freedom and culture of the mind," he explains. "For, in that, they think, lies
the happiness of life."[3] In Utopia, therefore, the goal is to
liberate people from the daily grind of their trades so they
have more time to improve themselves. It's akin to Karl
Marx's youthful vision, laid out in *The German Ideology*, of
creating a postrevolutionary society in which technology
will liberate us so that we can hunt in the morning, fish in
the afternoon, rear cattle in the evening, and criticize after
dinner. The aim is to nurture citizens who garden, exercise,
play music, and converse with one another. That's the real
work on More's imaginary island. The ultimate goal is to pay
people to do nothing except make themselves better citizens.
Utopia, then, is a kind of never-ending school designed to
improve its inhabitants. And More, as the author of *Utopia*, is a teacher too. He's raising consciousness among his
readers—encouraging us to imagine a place where one can
learn how to become a better human being.

Five hundred years after the publication of *Utopia*,
More's Renaissance humanism—with its focus on realizing
the "happiness of life," is back in vogue. It never went away
completely, of course. In the nineteenth century, a youthful
Karl Marx kept it alive. Today, however, rather than Utopia
or communism, it now goes under the name of "universal
basic income." This is the idea that, in our age of rising
technological unemployment and inequality, the government will give all its citizens—rich and poor, young and
old, male and female alike—a living wage whether or not
they have a job. "Money for Nothing" one headline about
universal basic income thus says.[4] "Sighing for Paradise to

Come" declares another about the future as a cornucopia of "technological abundance in which paid work is optional and no one goes without."[5]

Paradise or not, everyone today, it seems, both inside and outside Silicon Valley, is talking about universal basic income as the fix to the looming joblessness crisis of our smart machine age in which we will all become members of what Yuval Noah Harari calls the "useless class."[6] Its many proponents include libertarian technologists like the Y Combinator CEO Sam Altman, who is funding a trial in Oakland around it, as well as more traditional progressives such as the American labor organizer Andy Stern, the former president of the Service Employees International Union, who has written a book in favor of its implementation in the United States.[7] Local and national governments all over the globe—from Canada and Finland to Brazil, Holland, and Switzerland—are experimenting with referendums or pilot projects to reinvent the social security systems of the industrial age. In a world caught between the operating systems of the industrial and digital ages, it is, as the *Financial Times'* innovation editor John Thornhill summarizes—referring to a notion that, over the last five hundred years, has impressed thinkers as diverse as More, Thomas Paine, John Stuart Mill, Friedrich Hayek, and Milton Friedman—"an old idea with modern appeal."[8]

Given the central role of creative education in Utopia, it's not surprising that the Swiss political activist Daniel Straub, one of the world's most successful evangelists of universal basic income, is a former teacher. Straub once taught at a Zurich school that was based on the pedagogical ideas of the early-twentieth-century Italian educational reformer Maria Montessori. A critic of the rigid discipline and rote learning methods of traditional schools in industrial Italy, Montessori pioneered an education system that prioritized a child's self-initiative through creative work. Just as Thomas More believed in the value of hands-on learning, Montessori—who established her first school in Rome in 1907—believed that kids are best developed through the training of both their minds and their senses. Her revolutionary schools, therefore, did away with grade levels and desks and the traditional classroom, introducing a program of manual arts—gardening, keeping house in a model kitchen, gymnastics, and creative game-playing—designed to spontaneously nurture a lifelong self-discipline in the students. The system was originally dismissed by traditional educators as "utopian," but there are, today, some twenty thousand Montessori schools in 110 countries around the world, including five thousand in the United States. Two of the best-known Montessori graduates are Google cofounders Sergei Brin and Larry Page.

"It's appalling. Our schools are from the industrial age, from two hundred years ago," Straub, the father of a young child, complains to me about Switzerland's traditional schools, which, he says, encourage a stifling conformity in children. In contrast, his experience as a Montessori teacher confirmed his faith in human nature. In Montessori schools, he tells me, the kids aren't passive. Instead, he insists, "they have an inner drive to do things."

I'm meeting with Straub at his cramped first-floor office above a store that sells Christian iconography on the Augustinergasse, a winding cobbled lane in the medieval heart of Zurich. The old building, he tells me, built in 1365, was originally the home of the town's trumpet player. But music, he explains, was banned during the Reformation— and so the Augustinergasse house, like the rest of Zurich, remained silent for most of the sixteenth century.

During the Reformation, the great debate was about free will rather than universal basic income. All over sixteenth-century northern Europe—from England, Belgium, and Germany to Swiss cities such as Geneva, Basel, and Zurich—the conversation was about the role of human agency. It pitted Renaissance figures like Erasmus, More, and Holbein against populist preachers like Martin Luther and Huldrych Zwingli, the Swiss priest who initiated the Reformation in Zurich. It was a debate, you'll remember, between humanists who believed that we have freedom to shape our own histories and those religious fanatics who believed in the existentially disempowering idea of predestination. Today, half a millennium later, that debate has been reframed as one between technological determinists and those, like Edward Snowden, who believe in the primacy of human agency in determining the future.

Daniel Straub is himself an advertisement for the humanist argument that we do, indeed, have the freedom to shape the future. In 2007 he discovered that what he describes—in good Mitteleuropean fashion—as his "calling" was to increase "human consciousness" though the pursuit of universal basic income. "Most people see high unemployment as the problem," Straub tells me. But for him, smart machines are actually liberating. "Machines can free us to do what we want," he says, echoing Thomas More's argument in *Utopia* that leisure matters more than work. They can, he suggests, help us switch the music back on in our lives.

Technology, Straub argues, is dramatically altering the nature of work and employment. "My father had one job in his life. I have had six jobs. And my child will have six jobs at the same time," he explains. That's why, he thinks, we need the security of a basic income. It will give us what he calls a "platform" or a "foundation" to creatively navigate this turbulent new working environment.

And so Daniel Straub went to work on fixing the future. His goal, so to speak, was to transform Switzerland into a giant Montessori school, where everyone would be free to transform work into rigorous play. As in More's Utopia, Straub wanted to create the conditions for the "happiness of life" in Switzerland. In 2012 he and a small group of fellow activists began collecting signatures for a referendum on introducing an unconditional basic income pilot program. Under the Swiss constitution, to achieve the referendum they needed 100,000 signatures, which they collected by 2013. In the end, they got 120,000 people to sign the petition—2 percent of the entire Swiss adult population. In June 2016 the Swiss voted on a referendum, the first vote of its type in the world, requesting a basic monthly income of 2,500 Swiss francs (US $2,514) to all adults, and 625 francs (US $629) to every child.

You'd think, Thomas More says of Utopia's economy, that the "necessities of life would be in scant supply" in an economy where people work so little and where, indeed, there is neither private property nor money. But actually, he reassures us, Utopians have "more than enough of the necessities and even the conveniences of life" because everyone on the island contributes to the common good.[9] Straub's economics for his proposal are similar. To pay for the reform, he intended to redistribute a third of the annual Swiss GDP—with the other two-thirds being "untouchable." So, as in Utopia, everyone in Switzerland would contribute something toward the common good.

Although Straub lost the 2016 referendum with 77 percent of the Swiss voters going against his idea, he still sees in the vote a great victory. First, he reminds me, there were neighborhoods of Zurich in which a majority of the voters were in favor of the proposal. Second, and most important, he succeeded in raising *consciousness* about reinventing social security in the twenty-first century. Like Thomas More, Daniel Straub has put Utopia on the map. Before the referendum, he tells me, nobody in Switzerland had ever heard of a universal basic income. And now, he says, smiling broadly, "everyone has."

Anyway, Straub predicts, it's only a matter of time before we will be forced to implement some sort of radical reinvention of an industrial welfare system that is increasingly obsolescent. The problem, he explains, is that economic growth can't keep up with productivity. "There's no doubt in my mind," he thus says, "that at some point we will have basic minimum income."

Many other thinkers share Straub's confidence in the inevitable radical reinvention of our industrial social security system. From Zurich I fly to Amsterdam to meet with another of Europe's leading champions of the basic minimum income, Rutger Bregman. Originally from Utrecht, the Dutch city that is pioneering a 2017 scheme to pay its residents an unconditional monthly stipend, Bregman is the author of the aptly named *Utopia for Realists: How We Can Build the Ideal World*,[10] a bestselling polemic in favor of universal basic income, which has been translated into twenty languages. As we sit in a café outside Amsterdam Central Station on an unseasonably warm April afternoon, the youthful Bregman lays out the case for a universal basic income. Reiterating the core message in his book, that we need, in his words, to "control the future," Bregman—like Daniel Straub—celebrates the disappearance of what he calls "bullshit jobs" that dumb us down and drain us of our humanity. Above all, however, Bregman argues that we live in radically disruptive times that require what he calls "big ideas" if we are to successfully reinvent the industrial welfare system of benefits. The old left, in particular, he tells me, is intellectually bankrupt, which may explain why many traditional socialists, particularly in the labor unions, haven't embraced the universal basic income idea.

The case for a universal basic income is also shared by many technologists and entrepreneurs, who see it as an essential feature—perhaps even the central social security pillar—of tomorrow's networked society. Robin Chase, the former CEO of both Zipcar and Buzzcar and a leading evangelist for the sharing economy, tells me that universal basic income represents a kind of investment that will allow us to "tap into people's talents." There will, she promises me, be a "huge uptick in happiness, creativity, and productivity" after its introduction. More ominously, Stowe Boyd, a Boston-based influential technology commentator and researcher who describes himself as a "post-futurist," warns me that unless we introduce universal basic income, there will be "homelessness sixteen deep in the streets." Worse still, Boyd forecasts the eruption of a popular revolution he dubs the "Human Spring" by the mid-2020s if we don't address this issue now.

One of the world's leading authorities on technological unemployment is Martin Ford, the Silicon Valley–based author of the acclaimed *Rise of the Robots: Technology and the Threat of a Jobless Future*,[11] which won the *Financial Times'* business book of the year award in 2015. Over lunch at a Greek restaurant packed with tech workers in the Silicon Valley city of Sunnyvale, Ford tells me that he imagines the tipping point in terms of mass unemployment will come in fifteen or twenty years' time, with the crisis hitting low- and high-wage jobs in parallel, particularly such service center jobs as driver, store assistant, and office worker. "Nothing is off the table," Ford says, explaining the impact of smart technology on employment. "All the politicians talk about jobs coming back," he says bluntly, "but they aren't." And so, like everyone else, he believes that the "easiest, most realistic solution" is the universal basic income. He thinks it needs to "start small" in experimental pilot programs and be phased in gradually through the raising of broad-based taxes, including a carbon tax and a value-added tax (VAT). Like Robin Chase, Ford believes that this will encourage more people to be creative. Not everyone can become an entrepreneur, he tells me. "But if you give people safety nets," he says, "they'll take more risks."

Everywhere the message is the same. But few make the argument as idealistically as Albert Wenger, Brad

Burnham's partner at Union Square Ventures and—along with Burnham and John Borthwick—one of New York City's most prescient start-up technology investors. The Bavarian-born, American-educated Wenger makes universal basic income, which he calls "economic freedom," one of the three essential "freedoms" in what he calls, in his 2016 book, a *World After Capital*.[12] But giving people this monthly stipend isn't, in itself, the solution, Wenger says. It has to be accompanied by his other two freedoms. The second, "informational freedom," refers to a "re-decentralized" internet featuring peer-to-peer companies such as the micro-patronage network Patreon, which provides a platform for the buying and selling of creativity. And the third freedom is "psychological freedom," which is rooted in what he calls "self-regulation"—by which he means the ability to master oneself. So in Albert Wenger's utopia, "economic freedom" buys us the time to be creative; "psychological freedom" gives us the discipline to create; and "informational free-dom" gives us the operating system to buy and sell that creativity. What matters is that they exist in combination. Together they represent Wenger's ideal operating system for a "world after capital" in which, he promises, the "only scarcity" will be "our attention."

Like Daniel Straub, Wenger sees a reinvented education system as critical to fixing our more short-term future. This ability to master oneself, to balance the rational and emotional sides of the brain, he tells me when I visit him at the Union Square Ventures offices on Broadway in down-town New York City, is "superimportant," particularly on the internet, with its illusion of infinite freedom. Like Straub, he believes that the current educational system is mostly "broken" because it fails to teach children how to achieve this psychological freedom. And so his three teenage children—who are all homeschooled by Wenger and his wife, the serial tech entrepreneur Susan Danziger—spend their time reading books on Stoic philosophy, neuroplasticity, and Buddhism in preparation for their adult lives.

But homeschooling isn't practical for many working families, and very few parents offer the economic resources or intellectual firepower of Albert Wenger and Susan Danziger. So how can we fix the broken education system? And what, exactly, should we be teaching kids in order to prepare them for a life in which they might either be permanently unemployed—and relying on their monthly universal basic income stipend—or multitasking six jobs simultaneously?

What Are Humans Good For?

And so, finally, we've arrived at the question of education.

Education, we are told, particularly by those who don't teach or work in schools, is the answer. Education is how people are supposed to be retrained to work in the new economy. Education is where kids develop Albert Wenger's "psychological freedom" in order to break their online addictions. Education, to borrow some of Daniel Straub's language, is where we learn to be human. And though none of this is exactly wrong, the problem is that education has become the default solution to everything. When we don't know how to solve a big problem, we shove it into the classroom and make underpaid and overworked teachers responsible for fixing it. The bigger and more amorphous the problem, the more we hand it off to schools to fix.

So, for example, in their excellent bestselling *The Second Machine Age*—about the economic impact of Moore's Law on society—the MIT economists Erik Brynjolfsson and Andrew McAfee make education their number one policy recommendation for solving the American future. "Teach the children well," Brynjolfsson and McAfee conclude, speaking of the need to pay teachers better and make them more accountable for their pedagogical work, particularly the teaching of "hard-to-measure skills like creativity and unstructured problem-solving." Teachers should also, Brynjolfsson and McAfee advise, use new technology such as massive open online courses (MOOCs) to "enable low-cost replication of the best teachers, content, and methods."[13]

The truth, however, in America at least, is that the children aren't being taught well. A May 2017 Pew Research Center report, *The Future of Jobs and Jobs Training*, asked 1,408 senior American executives, college professors, and AI experts a series of questions about the challenges of educating people for an automated world. The Pew report found that 30 percent of them expressed "no confidence" that schools, universities, and job training will evolve sufficiently quickly to match the demand for workers over the next decade.[14] "Bosses believe your work skills will soon be useless," the *Washington Post* bluntly concluded about the report.[15]

"People are wrestling with this basic metaphysical question: What are humans good for?" Lee Rainie, the study's coauthor and the director of the Pew Research Center, said in response to the report. "It's important to figure that out because this blended world of machines and humans is already upon us and it's going to accelerate."[16]

So what, exactly, are humans good for? Especially when compared with the smart machines that, according to Pew's Lee Rainie, are "eating humans' jobs."[17]

I pose this question to Nicholas Carr, among America's most respected writers on the human costs of the digital revolution, the author of the Pulitzer Prize–nominated *The Shallows* and several other powerful books about technology. Over a dinner of Central Asian food at a funky Tajikistani teahouse in Boulder, Colorado, where Carr now lives, he talks metaphysics to me.

While he admits to being irritated when people label him a "humanist," he nonetheless offers a very vivid contrast between human beings and smart machines. "There's no gray zone in computers. They can't be ambivalent, and they can't be programmed to deal with ambiguity. Nor do they have intuition," he explains.

Like Stephen Wolfram who, you'll remember, denies that computers will ever have "goals," Carr believes that it's "highly implausible" that smart machines can become self-aware and acquire their own consciousness. "And the great thing about human consciousness in comparison with a robot," he adds, taking a swig of beer, "is that we can do different things simultaneously."

But how do we teach the children well? I ask him. What are the skills we should be teaching our kids so that they will have not only jobs but also relevance to what Pew's Lee Rainie calls the "blended world of machines and humans"?

Carr talks about Toyota's recent announcement that it is replacing robots with skilled craftsmen at some of its Japanese factories. Toyota has recognized that these workers have years and years of experience, which enables them to understand the constant ambiguity of the workplace. The same is true of the intuition that doctors have acquired over many years of physically working with their patients. None of this intuition, he explains, can be represented by an algorithm. Carr's focus is on what Matthew Crawford, another popular American writer, describes as "shop class as soulcraft."[18] Carr is suggesting, like Thomas More, that the uniquely human value of a trade lies in its practice. Education, then, Carr argues, sounding not unlike Maria Montessori, shouldn't just be about knowing; it's also about doing.

So that, according to Nicholas Carr, is what humans are *good for* in an age of increasingly smart machines. The challenge (and opportunity) for educators, then, is to teach everything that can't be replicated by a robot or an algorithm. For Carr, with his vision of the profound limitations of computers, that includes the nurturing of intuition, ambiguity, and self-awareness. For Daniel Straub, the former Montessori educator, it is the teaching of consciousness and the idea of a "calling." And for Union Square Ventures's Albert Wenger, who is homeschooling his three teenagers, it's teaching the self-mastery enabled by "psychological freedom."

This is the humanist ideal of education More laid out five hundred years ago in *Utopia*. It's the teaching of the unquantifiable: how to talk to one's peers, how to realize self-discipline, how to enjoy leisure, how to think independently, how to be a good citizen. But does this kind of innovative education really exist in the world today? Or is it, like More's imaginary island, a fictional idea without any grounding in reality?

Escaping the Silos

After my lunch with Martin Ford at the Greek restaurant in Sunnyvale, I drive a few miles north on US-101 to Palo Alto High School. This publicly funded school, situated over the road from Stanford University, lies in the epicenter of Silicon Valley, the risk-taking capital of the world. It is the high school to which tech titans like Steve Jobs send their kids to be educated.

In spite of that troubling 2017 Pew report about the poor quality of teaching in America, there fortunately remain some innovative schoolteachers who are successfully preparing their students for a future dominated by smart machines. I've come to Palo Alto High to talk to one of California's most acclaimed high school teachers about her pedagogical methods. Esther Wojcicki—or Woj, as she is popularly known—has taught journalism at Palo Alto High since 1984. She founded the school's 24,000-square-foot Media Arts Center, which today, with its six hundred students and nine publications, is the largest digital media program in the United States. She was selected as the California teacher of the year in 2002 and has won numerous other local and national teaching awards. And she is the coauthor of *Moonshots in Education*,[19] a 2014 book featuring a foreword by the actor James Franco, which promotes the idea of schools requiring students to invest 20 percent of their time in independent projects. Among her best-known students are Franco and Lisa Jobs, Steve Jobs's oldest daughter.

Wojcicki not only teaches the children of the tech aristocracy but is also the matriarch of one of Silicon Valley's most remarkable tech families. She is the mother of three adult daughters: Susan, Anne, and Janet. In 1998, when Larry Page and Sergei Brin founded Google as Stanford graduate students, Susan rented them space in her garage. Today Susan Wojcicki is the CEO of YouTube and among the most powerful entertainment moguls in the world. Her sister Janet is a professor of epidemiology at UC San Francisco. Anne, the youngest Wojcicki, is the cofounder and CEO of the gene mapping start-up 23andMe and was married to Google cofounder Sergei Brin. So Brin is Esther Wojcicki's former son-in-law, and they remain, she tells me, on very friendly terms.

We happen to meet on what one of Wojcicki's students referred to as "moonshot day"—the day of the week when the Media Arts Center provides students with its facilities and resources to do their own special projects. So the multimillion-dollar media center, lined with the latest iMacs, has the feel of a high-tech newsroom rather than a classroom. Every so often we are interrupted by a student asking Wojcicki a question. Her tone with them is edgy and challenging. She resembles as much a newspaper editor or a life coach as a conventional teacher. I watch Wojcicki interact with the students, simultaneously cajoling and encouraging them, forever trying to mine their talent. It occurs to me that her illustrious reputation as a high school teacher, with all the awards and acclaim, is based on her ability to reinvent herself as part friend, part mentor, part parent, and part boss to the students. The US managing editor of the *Financial Times*, Gillian Tett, in her influential 2015 book, *The Silo Effect*,[20] argues that the most successful people of the future will be those who escape from traditional institutional or vocational categories. Wojcicki, herself sidestepping the conventional pedagogical silos and treating her students as little adults, is also preparing them for a world where, as Daniel Straub predicts, people will do six different jobs simultaneously.

Not surprisingly, Wojcicki has very strong views about the failure of schools to educate kids for the twenty-first century. "We are still training kids for the twentieth century. The pedagogy is the same. Rather than training thinkers, we are training people who follow instructions. We are creating a nation of sheep," she says, referring to the teaching in most schools, including even that of some of her own colleagues at Palo Alto High. "We've got to stop telling kids what to do all the time. Instead, we need to treat them the way we ourselves would like to be treated."

In an age of AI, she says, it's particularly important for kids to "feel empowered" if they are to realize their potential. And so her goal as a teacher, she tells me, is to try to create technically proficient, confident kids who are willing to take risks and aren't afraid of failure. As we've seen so many times before in this book, it's all about trust. Trusting these kids is critical, she says. We have to *trust* them to make their own mistakes. "Any kids, average or otherwise," she says, "can learn from their mistakes."

That's how she brought up her girls, Wojcicki says. "I trusted my daughters with a lot of freedom," she confides. "I set a model. I was a rebel. They did what I did rather than what I said." In fact, Wojcicki boasts, the origins of some of Google's most influential ideas can be traced to her child-rearing strategies. So, for example, "Larry and Sergei got their twenty-percent rule from my daughters," she says, speaking of Google's corporate policy of requiring its staff to invest 20 percent of their time in what it calls "moonshot" projects, such as driverless cars, delivery drones, smart homes, and robots.[21]

But to her, Sergei Brin and Larry Page are also models of risk-takers unafraid of failure who are able to "think outside the box." "Both Sergei and Larry went to Montessori schools, so they've always set out to accomplish what everyone told them couldn't be done," Wojcicki, an admirer of Maria Montessori's teaching method, reminds me. So, in her mind, the innovative start-up entrepreneur and the successfully educated student actually share many of the same attributes: independence, a propensity to take risks, and a willingness to rethink conventional assumptions, take moonshots, and fix the future.

Returning to Utopia

If there is a country that is a successful model for educating its citizens, it's that tiny tropical island on the southern tip of Malacca that aspires to become the world's first Smart Nation. In dramatic contrast with the United States, Singapore is teaching its children well. Just as it's the most connected country in the world, the world's only city-state island also might be the best educated. In the latest Programme for International Student Assessment (PISA) study, an influential test conducted by the Organization for Economic Co-Operation and Development (OECD) every three years involving more than five hundred thousand students from more than seventy countries around the world, Singapore was ranked first overall in all three test subjects—math, reading, and science.[22] The United States, in contrast, didn't finish in the top twenty in any of the tests—limping in thirty-fifth in math, twenty-fifth in reading, and twenty-fourth in science.[23]

"Education is all we have" and *"People are our only resource"* are statements one hears often on an island where, as you'll remember, there are very few natural resources. In Singapore, 90 percent of its population of five million get some sort of post–high school education—with 25 percent of Singaporeans attending university, 40 percent attending polytechnic, and 25 percent going to one of the country's three Institutes of Technical Education (ITE) colleges. In many ways, these ITE colleges are Singapore's most notable achievement. The equivalent of Germany's much-vaunted apprenticeship system, the three ITE colleges offer two-year courses in digital media that are designed to give job skills to people who aren't academically gifted. The three colleges have more than fifty thousand students, almost all of them subsidized by the government, which pays between $500 and $900 a year for their education. Ninety percent of ITE graduates get a job within six months, many doing community service in Vietnam and Cambodia.

After spending an afternoon touring one of the cavernous ITE colleges and meeting its enthusiastic students, I talk with Bruce Poh, a former Hewlett-Packard engineer who is the CEO of the whole system. "Government must take leadership!" Poh exclaims, explaining the annual $470 million invested by the Singapore government in the three ITE colleges in order to create a digitally skilled workforce. "No skill equals no job—without education, people would otherwise end up on the street," he tells me, adding that the government's investment in collaborative robots, known as "cobots," will end up assisting, rather than replacing, workers in an increasingly automated age.

One American who knows the Singaporean education system intimately is Tom Magnanti, a former dean of the School of Engineering at MIT who has been living in Singapore for more than twenty years. Magnanti, who originally came to Singapore to run MIT's alliance with the National University of Singapore, is now the president of the Singapore University of Technology and Design (SUTD), an ambitious new $325 million university for thirteen hundred students, which opened in 2012. As a joint collaboration between MIT and the government, SUTD is known as "the new MIT in Singapore." It sits on the opposite end of the education spectrum from Singapore's ITE colleges. Rather than teaching students basic technical skills, the goal of Magnanti's important new university is to produce exceptional twenty-first-century leaders.

"I view myself as an academic entrepreneur," Magnanti tells me when we meet on SUTD's gleaming high-tech campus near Singapore's equally gleaming high-tech

Changi Airport. Rather than money, however, the value that Magnanti is trying to create as an academic entrepreneur is in human capital. As the founding president of SUTD, he is in the business of investing in technically literate and responsible leaders for the twenty-first century. Like that of Esther Wojcicki at Palo Alto High, Magnanti's goal is to realize the talent of his students. And like Wojcicki's Media Arts Center, Magnanti's new university is a moonshot project designed to reinvent the way we think about education in our network age.

One of his goals with SUTD, he explains, is to challenge the conventions of traditional technical universities, in terms of both its year-round academic calendar and its innovative curriculum. In its attempts to escape the academic silos of a typical technical university, SUTD doesn't offer traditional degrees in subjects like electrical or mechanical engineering. The interdisciplinary curriculum, instead, is built around the question "What products, services, and systems does the world need?" It is designed, according to Magnanti, to "create a sense of magic" in the value of not only technology but also the arts and humanities.

What the world definitely *needs*, Magnanti tells me, is leaders. That, indeed, is the real purpose of the new university. Quoting Benjamin Franklin's remark that the best leaders need to understand technology, he adds that they should understand everything else too. And that's his goal at SUTD: to "train citizens" to become the "leaders of the future" in a world where we are increasingly asking what humans are good for. Indeed, the university's mission statement is to "nurture technically grounded innovators to serve societal needs," and incoming students are selected according to portfolios, letters, and essays as much as their formal academic qualifications.

In contrast with MIT's Brynjolfsson and McAfee, Magnanti isn't a fan of MOOCs and online education. "Physical education isn't going away," the still trim Boston native—who confesses that he always wanted to play right field for the Red Sox—tells me. "I love these kids. I get such a thrill seeing them come here and leave three or four years later. That's what education is all about."

Magnanti's personal commitment to SUTD is both impressive and moving. With a severely disabled forty-five-year-old son in Boston, Magnanti commutes between Massachusetts and Singapore and has done the twenty-thousand-mile return journey more than a hundred times. Leaders beget leaders. Tom Magnanti is not only the most frequent flier on Team Human; he is also one of its most valuable members.

Lessons from More's Law

In Silicon Valley, the answer—at first glance, at least—to "the question of education" is simple: more technology. "Coding," Apple CEO Tim Cook told President Donald Trump about the best way to reform the American educational system, "should be a requirement in every public school."

Cook's vision is being realized by Code.org, a non-profit founded in 2012 by a couple of successful startup entrepreneurs, which has raised $60 million from Microsoft, Facebook, Google, and Salesforce. Code.org's goal is to make computer science as essential in the classroom as reading, writing, and arithmetic. Active now in twenty-four US states, Code.org offers free online classes that have been tried by 100 million students worldwide and has also provided training workshops to 57,000 teachers.[24]

But it's not just coding lessons that Big Tech wants to put into every school classroom. "In the space of just a few years," reports the *New York Times'* Natasha Singer, "technology giants have begun remaking the very nature of schooling on a vast scale, using some of the same techniques that have made their companies linchpins of the American economy." So, for example, Reid Hastings, the cofounder and CEO of Netflix, is championing an algorithmic math-teaching program called Dreambox in several US states including Texas, Maryland, and Virginia. Marc Benioff, the CEO of Salesforce, is handing out $100,000 "innovation grants" to middle school principals in order to transform their thinking from that of bureaucrats into start-up founders. And, in more than 100 US schools, Mark Zuckerberg is testing user-centric software partly built by Facebook that empowers kids to take control of their own learning and work with their teachers as mentors.[25]

There is no doubt, of course, that the often archaic education system needs to be radically reformed in today's digital age. That said, however, there are three major problems with this attempt by Big Tech to radically transform the traditional education system. The first is that the line between the public interest and the self-interest of these tech companies and individuals is often troublingly blurred, particularly in an education market that is supposed to reach

$21 billion by 2020. So, for example, Reid Hastings is not only an evangelist for the DreamBox software but is also what the *Times'* Natasha Singer calls the company's "guardian angel," having donated $11 million in 2009 to a nonprofit charter school to enable it to buy the algorithmic platform.

Second, the kind of philanthropy being practiced by Hastings, Benioff, and Zuckerberg is what Natasha Singer calls a "singular experiment in education, with millions of students serving as de facto beta testers for their ideas." The American education system is, essentially, being reinvented in Silicon Valley's image. So what Mark Benioff is trying to accomplish with those $100,000 "innovation grants" is a transformation of San Francisco schools into San Francisco start-ups. And what Mark Zuckerberg is doing, by testing that user-centric software in those 100 schools, is transforming the classroom into Facebook. "It looks more a Google or a Facebook than a school," noted the CEO of a charter-school network that is experimenting with Zuckerberg's software.[26]

But the biggest issue of all is whether the answer to the question of education really is more technology. It's all very well for other people's children to be educated by algorithms in Facebook-like classrooms, but when it comes to their own kids, then many of the most successful Silicon Valley entrepreneurs are much less evangelical about digital technology. As the New York University professor of psychology, Adam Alter, notes in his 2017 bestseller *Irresistible*, many of the "world's greatest public technocrats" —including Steve Jobs, Twitter cofounder Evan Williams, and former editor of *Wired* magazine Chris Anderson— "are also its greatest private technophobes."[27] Jobs, indeed, even discouraged his kids from using an iPad or iPhone.

In Zurich, I ask Daniel Straub—who, in addition to being a former Montessori teacher, used to work as a business development executive for IBM—about the role of digital technology in the Swiss educational crisis. Was it the problem or the solution? I ask. Do kids need more or less technology to develop into responsible adults?

"Technology is moving so fast, and we are stumbling behind," Straub tells me, sounding like another influential early-twentieth-century educational reformer, Rudolf Steiner, the father of Waldorf schools, the increasingly popular educational system that discourages the use of electronic screens. "It's not inherently human to sit on the couch and watch television or go on the internet all day."

Steiner was a turn-of-the-century Austrian philosopher and social reformer who invented an educational idea called "anthroposophy," a kind of metaphysical humanism influenced by Goethe and Nietzsche, which stressed the importance of the independent spirit, imagination, and will in a child's development. In 1919, in the wake of Germany's defeat in the First World War, Rudolf Steiner gave a lecture about anthroposophy at the Waldorf-Astoria cigarette factory in Stuttgart, which was owned by Emil Molt, one of his students. The lecture resulted in Molt's hiring Steiner to open a free public school, based upon anthroposophical ideas, for all the employees of his factory.

Although Steiner died in 1925, the Waldorf school movement, like its Montessori counterpart, is an increasingly popular alternative to traditional education. There are now more than a thousand independent Waldorf schools in sixty countries around the world. Like Montessori schools, the Waldorf teaching method focuses on nurturing the creative and moral development of young children. In contrast with Montessori's association of the idea of play with work, however, the Waldorf tradition focuses on the creative, emotional, and aesthetic development of the child. At Waldorf kindergarten schools, the teaching of reading and writing is put off till grade school, and as in More's Utopia, children focus on interactive games, music, creativity, and physical play. At grade school level, Waldorf schools stress what they call "social education," meaning the prioritizing of community values over individual competition.

In America, Waldorf education is best known for its controversial attitude toward technology, particularly screen media. Television and the internet are actively discouraged, if not banned, particularly in early grades. And much has been made of the growing popularity of Waldorf schools in the San Francisco Bay area, particularly in Silicon Valley, where senior executives from tech firms like Google, Apple, Yahoo, and Hewlett-Packard are known to send their kids to these schools.[28]

In an age of increasing internet addiction, online teen bullying, and social media narcissism, one reason for the increasing popularity of Waldorf education is its focus on the development of the child's self-control and social responsibility. The development of the "will"—meaning nurturing the determination, self-control, and commitment of the individual child—is central in Waldorf pedagogy. To say that Waldorf schools are teaching More's Law might be a slight exaggeration. But certainly the kind of duty-bound humanism that Thomas More preached in *Utopia* wouldn't be out of place in a Waldorf school.

At my dinner with Nicholas Carr in Colorado, I ask him if he thinks alternative educational traditions like Waldorf can help kids manage their online behavior more responsibly. Although digital media isn't the technological version of fat or sugar, he answers, echoing the arguments of the New York University psychologist Adam Alter in *Irresistible*, kids are finding it increasingly difficult to control their internet consumption. Ultimately, Carr acknowledges, the fix for a growing problem of online dependency and distraction has to come from users themselves rather than from the government. So yes, he says, the Waldorf model should be investigated by parents who are concerned about their children's irresponsible or uncontrollable use of digital media. Esther Wojcicki, however, is more ambivalent than Carr. While sympathetic to some of Waldorf's educational principles, the head of Palo Alto High's Media Arts Center is, unsurprisingly, strongly opposed to the banning of technology in the classroom.

The attitude I take toward Waldorf education, as a parent and as a critic, is similarly mixed. Both my kids, however, have attended Bay Area Waldorf schools. My son, as curmudgeonly as his father, found the focus on handwork, singing, and "eurythmy" (Waldorf's metaphysical version of calisthenics) intolerable and left as soon as he could. I also failed dismally as a Waldorf parent. At one parent meeting for my son's sixth-grade class we were asked about our home "media policy." As we went around the room, parent after parent boasted about not allowing the kids access to any media at home. One typically earnest Northern California couple confessed that they sometimes let their kids listen to classical music on the radio. When it was my turn to speak, I told the group that I actively encouraged my kids to watch movies on one of the many screens in my house. None of the other parents ever spoke to me again.

But my academically gifted and highly motivated teenage daughter has thrived at Waldorf and currently attends the Credo High School in the nondescript town of Rohnert Park, some fifty miles north of San Francisco. Credo, which is now America's second-largest public Waldorf high school, was founded by Chip Romer, a onetime aspiring writer and co-owner of a comedy club, who originally became involved in the Waldorf movement as a parent. We meet in the school's new campus in a sprawling industrial estate on the edge of Rohnert Park. In contrast with Palo Alto High, there are no tech-rich new media complexes on the Credo site. And unlike Tom Magnanti's gleaming new university in Singapore, Credo isn't supported by hundreds of millions of dollars of government investment.

Romer tells me that learning at the school is "rigorous" but "meaningful" and that 94 percent of Credo's graduating seniors meet the academic threshold to apply to the intensely competitive University of California system. At the conventional local public high school, in contrast, only 20 percent of students qualify. The pedagogical focus, he says, is on the development of "creative thinkers" in preparation for what he calls a "freelance society"—Daniel Straub's world in which we have six simultaneous jobs.

We talk about the idea of "will" in Waldorf philosophy. I ask if it also might translate as "agency," that word that has been featured throughout this book—from Edward Snowden's warning on the big screen in Berlin's Alte Teppichfabrik to the many members of Team Human determinedly working to shape technology for the benefit of their communities.

"Yes, agency is a good word to describe our philosophy," he confirms. "It's about building the capacity to achieve whatever you want. But it also comes with a sense of responsibility—particularly with being kind and respectful to others. Kids who struggle here have a lack of agency."

"So where does agency come from?"

"It comes from practice and experience," he replies. "It's like a muscle. The more you use it, the stronger it becomes."

In response to the "What are humans good for?" question, I suspect that Chip Romer would say that they are good at being good—at growing up into responsible citizens and caring people. That's also the goal of a Waldorf education, in which each child is treated as a discrete moonshot. And so, if we are to solve the plethora of challenges on the horizon—from wide-scale unemployment to the alienation and loneliness created by the always-on nature of our social media culture—we could do a lot worse than support alternative public schools like Credo. Waldorf schools certainly aren't perfect—and, as I warned earlier, education shouldn't be considered the panacea for our digital ills. But given that there are no magical solutions to all these problems, a humanist school like Credo—with its utopian curriculum—is a perfectly imperfect place to begin to address how we can fix the future.

CONCLUSION

OUR KIDS

"Nineteenth-century civilization has collapsed," warned Karl Polanyi at the beginning of *The Great Transformation*, his analysis of the shift from an agricultural to an industrial market society.[1] Today, the next chapter of this great technological transformation is taking place—the early-twenty-first-century transition from an industrial to a networked operating system, powered by the internet. And today, it's twentieth-century civilization that appears, in some ways, to be collapsing.

We have seen how the deeply disruptive changes wrought by the internet—with its new smart machinery, new modes of production, new types of economic scarcity and abundance, new models of work, new forms of culture, new educational systems, new inequalities and injustices, new definitions of wealth and poverty—are bewildering for those still clinging to twentieth-century realities. This is, indeed, the great transformation of our age, a complex journey that we are all navigating.

I confess that the fixes in this book, like their human agents, aren't entirely perfect. The big data innovations in Estonia and Singapore will understandably make traditional nineteenth-century-style liberals—with their belief in an individual's intrinsic right to privacy—nervous. Even responsible regulators like Margrethe Vestager are occasionally prone to overregulate. Some of the innovative solutions we've outlined, like blocking online advertising, come with a whole new set of problems. A labor strike led by multimillionaire artists like Taylor Swift is obviously not a little absurd. Educating "useless" people for a world without work is, in some ways, a Sisyphean task. Relying on the super-citizens of our gilded age like Jeff Bezos or Mark Zuckerberg to save the world is, to say the least, risky. And paying people to do nothing in an age of smart machines—as idealists like Rutger Bregman and the Union Square Ventures partner Albert Wenger want—is not going to automatically create a Marxian idyll of farmers, fishermen, hunters, and critics.

In contrast with Holbein's map of the little island of Utopia, our map is global. In the mid-nineteenth century, the economic historian Eric Hobsbawn reminds us, Britain was the only genuinely industrialized country. Fifty years later, Germany and the United States had overtaken it, and today almost every country in the world has experienced its own industrial revolution. Much the same is true today of the digital revolution. True, it all began in Silicon Valley, that narrow peninsula of Northern Californian land between San Francisco and San Jose that has spawned Hewlett-Packard, Apple, Google, Facebook, Uber, and Airbnb. And true, the first fifty years of our digital history have taken place on

Silicon Valley terms in what Karl Polanyi, critiquing the industrial revolution, called a utopian self-regulated market. But this libertarian model isn't sustainable. As we've shown, much of the real digital innovation in everything from new business models to regulation to ethics is now taking place away from the private superpowers of Silicon Valley—in Singapore, Estonia, Germany, India, China, even Oakland.

Our map is a multidimensional one. Our five tools for building the map of the future—regulation, innovation, social responsibility, worker and consumer choice, and education—are all essential building blocks of the future. But they aren't cut off from one another. As we've shown, there is no innovation without regulation. And as my conversations with socially responsible venture capitalists like John Borthwick, Freada Kapor Klein, and Steve Case demonstrate, successful innovators and wealthy investors can—and should—be good citizens too. This combinatorial architecture is critical to fixing the future. Government regulation and free market innovation aren't, in themselves, sufficient solutions. Their value lies in being combined, consciously or otherwise, with our other tools.

The new geography outlined on our map is also an old geography. Certainly nothing ever really changes in terms of the way in which dramatic disruption requires humans to assert their agency in order to fix the future. Pretty much everything about today's digital revolution has, one way or another, happened before. Even the Moore's Law versus More's Law debate, which boils down to the question of whether humans have real agency, is a recurring discussion throughout history. It preoccupied nineteenth-century critics of capitalism, pitting deterministic Marxist-Leninists, with their faith in the *inevitability* of history, against socialist humanists like Rosa Luxemburg. And it was the most controversial theological issue of the sixteenth century, dividing Renaissance humanists such as Erasmus and Thomas More from Martin Luther and other Reformation preachers.

What I hope this book has shown is that designing a coherent map of the future requires a coherent knowledge of the past. That's why knowledge of the history of humanism is so valuable in an age in which we are once again grappling with the eternal issue of what it means to be human. That's why the history of the nineteenth-century industrial revolution is so instructive in helping us make sense of the early digital twentieth-first century. And that's why a familiarity with the modern history of everything from robber baron capitalism to the contemporary automobile and food industries can be so useful today in determining how best to reform the most troubling aspects of the digital revolution.

But there is still a final question, one last thing, to answer about the map drawn in this book. Just as Thomas More and his humanist friends created a map of Utopia that, when looked at in a certain way, reveals something quite unexpected—a human skull—we too need to reimagine our map's message.

So, if you step back from the map laid out in this people's book and look at it with fresh eyes, what—or whom—should you see?

Rather than a skull, what you should see is the image of a young person. I dedicated this book about fixing the future *"For our kids."* But this might have been a little patronizing.

"*For* our kids" assumes that they lack agency of their own and that "the future" is something finished—something we adults will donate to them once it has been fixed.

But the good news is that our kids—as activists, entrepreneurs, and consumers—are already actively shaping this future. As we saw, for example, at the Kapor Center hackathon in Oakland, it's young people—like Tiffany Smith, the MBA graduate student from Chicago with the app for empowering the previously incarcerated—who are using technology for the social good. These kinds of hackathons are becoming increasingly popular among young technologists and social entrepreneurs. Indeed, the San Francisco–based technology nonprofit group Code for America even now organizes an annual CodeAcross hackathon day for the social good; it is held simultaneously in numerous countries around the world, including Brazil, South Korea, Pakistan, and Croatia. That "muscle," which Credo principal Chip Romer says is essential to good citizenship, is being strengthened by young people's commitment to technology for the good. "The more you use it," you'll remember Romer saying of this muscle, "the stronger it becomes."

Much has been made of young people as "digital natives"—the vanguard of the network revolution.[2] But it's no coincidence that these sentimental commentaries tend to be written by older people who have a romanticized, Rousseauian faith in the supposedly intuitive literacy of our kids. The truth, however, is that it is our kids who are the grizzled realists of all things digital, especially as consumers. It's these kids, for example, who are pioneering the "revenge of analog":[3] a return to the future of handwritten notebooks, printed books, and vinyl records—the "new" billion-dollar music business that now generates more profit than the entire streaming industry.[4] This isn't a Luddite moment. Young people certainly aren't smashing their digital devices. Instead, as consumers, they are discovering that some meaningful things—like listening to high-quality music recordings and reading or writing complex ideas—are most enjoyably done physically rather than virtually. Indeed, they are even beginning to pay again for their internet content, with the Reuters Institute reporting that the proportion of people aged between eighteen and twenty-four in the United States paying for their online news rose from 4 percent in 2016 to 18 percent in 2017.[5]

Ultimately, too, a solution to the great cultural problems of the networked age—digital overconsumption, distraction, and, above all, addiction—will also originate from our kids. As Nicholas Carr told me, there's never been a countercultural movement in history that hasn't come from young people. Digital culture—especially obsessive social media usage—has become so stultifyingly orthodox that it will inevitably be rejected by future generations of free-thinking kids. Just as the kids of the sixties rebelled against the cloying consumerism of late-twentieth-century capitalist society, so too, I suspect, we are now on the brink of a youthful rebellion against an increasingly conformist digital lifestyle of endless tweets, Facebook updates, and Snapchats.

Rather than rejecting digital culture, however, the most important challenge for young people is getting our new twenty-first-century operating system to work properly. The new networked world that we are now navigating is, in some ways, profoundly foreign to the industrial one. It's an unfamiliar map. For example, new divisions between nationalists and globalists have replaced the old conservative versus liberal political debates of the twentieth century, thereby signaling the end of the "left" and the "right" as we know them.[6] Young people are on the progressive side of this debate—voting disproportionately against Trump and Brexit. In the UK, 75 percent of eighteen- to twenty-four-year-olds voted to stay in the EU.[7] In the United States, Hillary Clinton won 55 percent of the millennial electorate.[8]

The digital transformation of the world is central to this new twenty-first-century debate between nationalists and globalists. As the *Financial Times'* John Thornhill predicts, "the more internationalist-minded tech industry may well act as a counterweight to the increasingly inward-looking nativism seen in the US and the UK."[9] But, as we've seen in this book, the direction of this new technocentric internationalism will come as much from Germany, Estonia, India, and Singapore as it will from Silicon Valley.

And all this change is still in its very earliest stages. It's 1850 all over again. A familiarly unfamiliar moment. "Nothing, nothing, nothing": that's how Brad Burnham described this kind of epochal change in which a whole generation will reinvent the world. The next chapter in our human narrative. "And then something dramatic."

ACKNOWLEDGMENTS

Just as the future can be fixed only combinatorially, so too this book is the fruit of a combination of folks. It was conceived by the Grove Atlantic publisher Morgan Entrekin and my agent Toby Mundy, who came up with its overall structure, historical themes, and title. It was expertly edited by Grove Atlantic's ever-patient Peter Blackstock and copyedited by the eagle-eyed Julia Berner-Tobin and Maxine Bartow. It was meticulously researched by Lyn Davidson at San Francisco's Mechanics' Institute Library. And its ideas came from the many people around the world who have been generous enough to meet with me. All I've done is joined up the dots of their wisdom. As I suggested in the introduction, this is as much their map of the future as mine.

NOTES

Epigraph

1. Thomas More, *Utopia*, eds. George M. Logan and Robert M. Adams (Cambridge University Press, 1975), 48.

Introduction

1. "Digital Transformation of Industries: Demystifying Digital and Securing $100 Trillion for Society and Industry by 2025," World Economic Forum, January 2016.

2. "System Crash," *Economist*, November 12, 2017.

3. Kevin Kelly, *The Inevitable: Understanding the 12 Technological Forces That Will Shape Our Future* (Viking, 2016)

Chapter One

1. For more on Wiener, Bush, and Licklider's role in the creation of the internet, see Andrew Keen, *The Internet Is Not the Answer* (Grove Atlantic, 2015), 14–18.

2. For an excellent overview of the nineteenth-century origins of privacy, see: Jill Lepore, "The Prism: Privacy in an Age of Publicity," *New Yorker*, June 24, 2013.

3. Samuel Warren and Louis Brandeis, "The Right to Privacy," *Harvard Law Review*, vol. 4, no. 5, December 15, 1890.

4. *Electronics Magazine*, vol. 38, no. 8, April 19, 1965.

5. Gordon Moore didn't himself call it Moore's Law when he wrote his 1965 paper. His friend Carver Mead was the first to call it that, in 1975.

6. Thomas L. Friedman, *Thanks You for Being Late: An Optimist's Guide to Thriving in the Age of Acceleration* (Farrar, Straus and Giroux, 2016), 27.

7. Ibid., 28.

8. Joi Ito and Jeff Howe, *Whiplash: How to Survive Our Faster Future* (Grand Central Publishing, 2016).

9. Friedman, *Thank You for Being Late*, 4.

10. Dov Seidman, "From the Knowledge Economy to the Human Economy," *Harvard Review*, November 12, 2014.

11. "2017 Edelman TRUST BAROMETER Reveals Global Implosion of Trust," Edelman.com, January 15, 2017.

12. Akash Kapur, "Utopia Makes a Comeback," *New Yorker*, October 3, 2016.

13. Oscar Wilde, "The Soul of Man Under Socialism" (1891), in Wilde, *The Soul of Man Under Socialism and Selected Critical Prose*, ed., Linda C. Dowling (London, 2001), 41.

14. More, *Utopia*, xxx.

15. For an excellent introduction of Holbein's work, particularly *The Ambassadors*, as it relates to both humanism and the Renaissance, see John Carroll, *Humanism: The Wreck of Western Culture* (Fontana, 1993), 27–35.

16. A dentist uncovered this peculiar subtext to More's map in 2005. For more on this unusual discovery, see Ashley Baynton-Williams, *The Curious Map Book* (University of Chicago Press, 2015), 14–15.

17. Friedman, *Thank You for Being Late*, 36–84.

18. Gerd Leonard, *Technology Versus Humanity: The Coming Clash Between Man and Machine* (Fast Future, 2016).

19. Richard Watson, *Digital Versus Human: How We'll Live, Love and Think in the Future* (Scribe, 2016).

20. Yuval Noah Harari, "Yuval Noah Harari on Big Data, Google and the End of Free Will," *Financial Times*, August 26, 2016.

21. Richard Metzger, "Capitalism's Operating System Has Gone Off the Rails: An Interview with Douglas Rushkoff," *Dangerousminds*, March 8, 2016 (http://dangerousminds.net/comments/capitalisms_operating_system_has_gone_off_the_rails_an_interview_with_dougl).

22. Klaus Schwab, "How Can We Embrace the Opportunities of the Fourth Industrial Revolution," WEForum.org, January 15, 2015.

23. Ibid.

24. Stephen Wolfram, *Idea Makers: Personal Perspectives on the Lies & Ideas of Some Notable People* (Wolfram Publishing, 2016), 78.

Chapter Two

1. "Knight Foundation, Omidyar Network and LinkedIn Founder Reid Hoffman Create $27 Million Fund to Research Artificial Intelligence for the Public Interest," Knightfoundation.org, January 10, 2017.

2. Amy Harmon, "AOL Official and Lawyer for Microsoft Spar in Court," *New York Times*, April 5, 2002.

3. John Borthwick and Jeff Jarvis, "A Call for Cooperation Against Fake News," *Medium*, November 18, 2016.

4. Eric Hobsbawm, *The Age of Revolution 1789–1848* (Vintage, 1996), 10.

5. Eric Hobsbawm, *The Age of Capital, 1848–1875* (Vintage, 1996), 39.

6. Ibid., 40.

7. Karl Marx and Friedrich Engels, *The Communist Manifesto* (Penguin, 2006), 7.

8. Ibid.

9. Karl Polanyi, *The Great Transformation: The Political and Economic Origins of Our Time* (Beacon Press, 2001), 41.

10. Ibid., 106.

11. More, *Utopia*, 18.

12. Polanyi, *The Great Transformation*, 35.

13. Ibid., 3.

14. Hobsbawm, *The Age of Revolution*, 202.

15. Ibid., 204.

16. Hobsbawm, *The Age of Capital*, 221.

17. Christine Meisner Rosen, "The Role of Pollution and Litigation in the Development of the U.S. Meatpacking Industry, 1865–1880," *Enterprise & Society*, June 2007.

18. Ibid., 212.

19. Marx and Engels, *The Communist Manifesto*, 32–33.

20. Carl Benedikt Frey and Michael A. Osborne, "The Future of Employment: How Susceptible Are Jobs to Computerization," Oxford Martin School, September 2013.

21. Steve Lohr, "Robots Will Take Jobs, but Not as Fast as Some Fear, New Report Says," *New York Times*, January 12, 2017.

22. Friedrich Engels, *The Condition of the Working-Class in England in 1844* (Swan Sonnenschein, 1892), 70.

Chapter Three

1. "How Did Estonia Become a Leader in Technology?" *Economist*, July 30, 2013. See also Romain Gueugneau, "Estonia, How a Former Soviet State Became the Next Silicon Valley," *Worldcrunch.com*, February 25, 2013.

2. Chris Williams, "AI Guru Ng: Fearing a Rise of Killer Robots Is Like Worrying About Overpopulation on Mars," *Register*, March 19, 2015.

3. *Times*, December 23, 2016.

4. " 'Irresistible' by Design: It's No Accident You Can't Stop Looking at the Screen," *All Tech Considered*, NPR, March 13, 2017.

5. See David Streitfeld, " 'The Internet Is Broken': @ev Is Trying to Salvage It," *New York Times*, May 20, 2017. Also Anthony Cuthbertson, "Wikipedia Founder Jimmy Wales Believes He Can Fix Fake News with Wikitribune Product," *Newsweek*, April 25, 2017.

6. Astra Taylor, *The People's Platform: Taking Back Power and Culture in the Digital Age* (Metropolitan Books, 2014).

7. Jaron Lanier, *Who Owns the Future?* (Simon & Schuster, 2013), 336.

8. Keen, *The Internet Is Not the Answer*, 27–28.

9. Ibid., 182.

10. Geoff Descreumaux, "One Minute on the Internet in 2016," *wersm .com*, April 22, 2016.

11. The idea that "data is the new oil" has been expressed by numerous pundits including Meglena Kuneva, the European consumer commissioner; the Silicon Valley venture capitalist Ann Winblad; and the IBM CEO Virginia Rometty.

12. John Gapper, "LinkedIn Swaps Business Cards with Microsoft," *Financial Times*, June 15, 2016.

13. Quentin Hardy, "The Web's Creator Looks to Reinvent It," *New York Times*, June 7, 2016.

14. Rana Foroohar, "Tech 'Superstars' Risk a Populist Backlash," *Financial Times*, April 23, 2017.

15. Keen, *The Internet Is Not the Answer*.

16. Ibid., 43.

17. Ibid.

18. Hannah Kuchler, "Google Set to Introduce Adverts on Map Service," *Financial Times*, May 24, 2016.

19. According to Gartner, 86.2% of global smartphone purchases in the second quarter of 2016 were for phones operating on the Android platform. See: Natasha Lomas, "Android's Smartphone Marketshare Hit 86.2% in Q2," *Techcrunch*, August 18, 2016.

20. Jerry Brotton, *The History of the World in 12 Maps* (Viking Penguin, 2012), 431–32.

21. Jonathan Taplin, *Move Fast and Break Things: How Facebook, Google and Amazon Cornered Culture and Undermined Democracy* (Little Brown, 2017), 4.

22. John Gapper, "YouTube Is Big Enough to Take Responsibility for Piracy," *Financial Times*, May 19, 2016.

23. Emily Bell, "Facebook Is Eating the World," *Columbia Journalism Review*, March 7, 2016.

24. Ibid.

25. Margot E. Kaminski and Kate Klonick, "Facebook, Free Expression and the Power of a Leak," *New York Times*, June 27, 2017.

26. John Herrman, "Media Websites Battle Faltering Ad Revenue and Traffic," *New York Times*, April 17, 2016.

27. Eli Pariser, *The Filter Bubble: What the Internet Is Hiding from You* (Penguin, 2011).

28. "2017 Edelman TRUST BAROMETER," Edelman.com, January 15, 3017.

29. Allister Heath, "Fake News Is Killing People's Minds, Says Apple Boss Tim Cook," *Telegraph*, February 10, 2017.

30. Nir Eyal, *Hooked: How to Build Habit-Forming Products* (Portfolio, 2014).

31. Andrew Sullivan, "I Used to Be a Human Being," *New York Magazine*, September 18, 2016.

32. Adam Alter, *Irresistible: The Rise of Addictive Technology and the Business of Keeping Us Hooked* (Penguin, 2017), 4.

33. Izabella Kaminska, "Our Digital Addiction Is Making Us Miserable," by *Financial Times*, July 5, 2017.

34. Bianca Bosker, "The Binge Breaker," *Atlantic*, October 8, 2016.

35. Saleha Mohsin, "Silicon Valley Cozies Up to Washington, Outspending Wall Street 2–1," *Bloomberg*, October 23, 2016.

36. This is calculated on August 6, 2016, capitalization numbers, which value Apple at $579 billion, Alphabet at $543 billion, Microsoft at $454 billion, Amazon at $366 billion, and Facebook at $364 billion. Collectively they were worth $2.306 trillion. In 2015, the Indian economy, the eighth-largest in the world, had a nominal GDP of $2.09 trillion.

37. David Curran, "These 9 Bay Area Billionaires Have the Same Net Worth as 1.8 Billion People," *SFGate*, February 1, 2017.

38. Alastair Gee, "More Than One-Third of Schoolchildren Are Homeless in Shadow of Silicon Valley," *Guardian*, December 28, 2016.

39. Jeffrey D. Sachs, "Smart Machines and the Future of Jobs," *Boston Globe*, October 10, 2016.

40. Paul Lewis, "California's Would-Be Governor Prepares for Battle Against Job-Killing Robots," *Guardian*, June 5, 2017.

41. "Digital Technologies: Huge Development Potential Remains Out of Sight for the Four Billion Who Lack Internet Access," Worldbank .org, January 13, 2016.

42. Ibid.

43. Somini Sengupta, "Internet May Be Widening Inequality, Report Says," *New York Times*, January 14, 2016.

Chapter Four

1. Robert D. Kaplan, *The Revenge of Geography* (Random House, 2012).

2. It was a remark made in 1970 by the American geographer Waldo Tobler. See Brotton, *The History of the World in 12 Maps*, 428.

3. Mark Scott, "Estonians Embrace Life in a Digital World," *New York Times*, October 8, 2014.

4. Tim Mansel, "How Estonia Became E-stonia," *BBC News*, May 16, 2014.

5. Sten Tamkivi, "Lessons from the World's Most Tech-Savvy Government," *Atlantic*, January 24, 2014.

6. Alec Ross, *The Industries of the Future* (Simon & Schuster, 2016), 5.

7. Ibid., 208.

8. https://e-estonia.com/facts.

9. Press release.

10. "Estonians' Trust in Parliament, Government Much Higher Than EU Average," *Baltic Times*, December 29, 2014.

11. "Linnar Viik—Estonia's Mr. Internet," *EUbusiness.com*, April 20, 2004.

12. Don Tapscott and Alex Tapscott, *Blockchain Revolution: How the Technology Behind Bitcoin Is Changing Money, Business, and the World* (Portfolio, 2016), 6.

13. Ibid.

14. More, *Utopia*, 46.

15. Ibid., 79.

16. Jeremy Rifkin, *The End of Work: The Decline of the Global Labor Force and the Down of the Post-Market Era* (Tarcher, 1996).

17. Andreas Weigend, *Data for the People: How to Make Our Post-Privacy Economy Work for You* (Basic, 2017).

18. Patrick Howell O'Neill, "The Cyberattack That Changed the World," *Daily Dot*, February 24, 2017.

19. Peter Pomerantsev, "The Hidden Author of Putinism," *Atlantic*, November 7, 2014.

20. Peter Pomerantsev, "Russia and the Menace of Unreality," *Atlantic*, September 9, 2014.

21. Shaun Walker, "Salutin' Putin: Inside a Russian Troll House," *Guardian*, April 2, 2015.

22. Andrew E. Kramer, "How Russia Recruited Elite Hackers for Its Cyberwar," *New York Times*, December 29, 2016.

23. Sam Jones, "Russia's Cyber Warriors," *Financial Times*, February 24, 2017.

24. Mark Scott and Melissa Eddy, "Europe Combats a New Foe of Political Stability: Fake News," *New York Times*, February 20, 2017.

25. For a compelling narrative of the Aadhaar project, see Nandan Nilekani and Viral Shah, *Rebooting India: Realizing a Billion Aspirations* (Penguin, 2015).

Chapter Five

1. James Manyika, Susan Lund, Jacques Bughin, Jonathan Woetzel, Kalin Stamenov, and Dhruv Dhingra, "Digital Globalization: The New Era of Global Flows," Mckinsey.com, February 2016.

2. Rodolphe De Koninck, Julie Drolet, and Marc Girard, *Singapore: An Atlas of Perpetual Territorial Transformation* (National University of Singapore Press, 2008), 14.

3. Jake Maxwell Watts and Newley Purnell, "Singapore Is Taking the 'Smart City' to a Whole New Level," *Wall Street Journal*, April 24, 2016.

4. More, *Utopia*, 59.

5. Watts and Purnell, "Singapore Is Taking the 'Smart City' to a Whole New Level."

6. "The Government of Singapore Says It Welcomes Criticism, but Its Critics Still Suffer," *Economist*, March 9, 2017.

7. Ibid.

8. Amnesty International, "Singapore: Government Critics, Bloggers and Human Rights Defenders Penalized for Speaking Out," *Online Citizen*, June 19, 2016.

9. Ishaan Tharoor, "What Lee Kuan Yew Got Wrong About Asia," *Washington Post*, March 23, 2015.

10. More, *Utopia*, 42.

11. Daniel Tencer, "Richest Countries in the World 2050: Singapore Wins, U.S. and Canada Hang in There," *Huffington Post*, November 6, 2012.

12. Sean Gallagher, "Prime Minister of Singapore Shares His C++ Code for Sudoku Solver," *ArsTechnica UK*, May 4, 2015.

13. "Trust Between Citizens, Government Key for Smart Nation: PM Lee Hsien Loong," *Straits Times*, July 12, 2016.

14. Watts and Purnell, "Singapore Is Taking the 'Smart City' to a Whole New Level."

15. "China's Tech Trailblazers," *Economist*, August 6, 2016.

16. "WeChat's World," *Economist*, August 6, 2016.

17. John Naughton, "The Secret Army of Cheerleaders Policing China's Internet," *Guardian*, May 29, 2016.

18. Marlow Stern, " 'Web Junkie' Is a Harrowing Documentary on China's Internet Addiction Rehab Clinics," *Daily Beast*, January 20, 2014.

19. "Creating a Digital Totalitarian State," *Economist*, December 17, 2016.

20. Simon Denyer, "China's Plan to Organize Its Society Relies on 'Big Data' to Rate Everyone," *Washington Post*, October 22, 2016.

21. Josh Chin and Liza Lin, "China's All-Seeing Surveillance State Is Reading Its Citizens' Faces," *Wall Street Journal*, June 26, 2017.

22. "Creating a Digital Totalitarian State."

23. Josh Chin and Gillian Wong, "China's New Tool for Social Control: A Credit Rating for Everything," *Wall Street Journal*, November 28, 2016.

24. "Creating a Digital Totalitarian State."

25. Celia Hatton, "China 'Social Credit': Beijing Sets Up Huge System," *BBC News*, October 26, 2015.

Chapter Six

1. Philip Stephens, "How to Save Capitalism from Capitalists," *Financial Times*, September 14, 2016.

2. Michael Tavel Clarke, *These Days of Large Things: The Culture of Size in America, 1865–1930* (University of Michigan Press, 2007).

3. This choice between democracy or oligarchy in today's age of digital inequality is once again a familiar theme, particularly with critics of free market capitalism. See, for example: Joel Kotkin, "Amazon Eats Up Whole Foods as the New Masters of the Universe Plunder America," *Daily Beast*, June 19, 2017.

4. Stephens, "How to Save Capitalism from Capitalists."

5. Farhad Manjoo, "Tech Giants Seem Invincible. That Worries Lawmakers," *New York Times*, January 4, 2017.

6. Philip Stephens, "Europe Rewrites the Rules for Silicon Valley," *Financial Times*, November 2, 2016.

7. Scott Malcomson, *Splinternet: How Geopolitics and Commerce Are Fragmenting the World Wide Web* (OR Books, 2016).

8. Farhad Manjoo, "Why the World Is Drawing Battle Lines Against American Tech Giants," *New York Times*, June 1, 2016.

9. Aoife White, "EU's Vestager Considers Third Antitrust Case Against Google," *Bloomberg*, May 13, 2016.

10. Murad Ahmed, "Obama Attacks Europe over Technology Protectionism," *Financial Times*, February 16, 2015.

11. Murad Ahmed, "Here's Exactly How Dominant Google Is in Europe in Search, Smartphones, and Browsers," *Business Insider*, April 20, 2016.

12. Mark Scott, "Phone Makers Key in Google Case," *New York Times*, April 20, 2016.

13. Samuel Gibbs, "Google Dismisses European Commission Shopping Charges as 'Wrong,'" *Guardian*, November 3, 2016.

14. Daniel Boffery, "Google Fined Record 2.4 Billion Euros by EU over Search Engine Results," *Guardian*, June 27, 2017.

15. John Naughton, "Challenges to Silicon Valley Won't Just Come from Europe," *Guardian*, July 2, 2017.

16. Scott, "Phone Makers Key in Google Case."

17. Conor Dougherty, "Courtroom Warrior Opens European Front in His Battle with Tech Giants," *New York Times*, September 29, 2015.

18. Peter B. Doran, *Breaking Rockerfeller: The Incredible Story of the Ambitious Rivals Who Toppled an Oil Empire* (Viking, 2016)

19. Gary L. Reback, *Free the Market! Why Only Government Can Keep the Marketplace Competitive* (Portfolio, 2009).

20. opensecrets.org.

21. Sally Hubbard, "Amazon and Google May Face Antitrust Scrutiny Under Trump," *Forbes*, February 8, 2017.

22. "The World's Most Valuable Resource," *Economist*, May 6, 2017.

23. "Amazon's empire," *Economist*, March 25, 2017

24. Catherine Boyle, "Clinton and Sanders: Why the Big Deal About Denmark?" *CNBC.com*, October 14, 2015.

25. Rochelle Toplensky, "A Career That Inspired 'Borgen,'" *Financial Times*, December 8, 2016.

26. Ibid.

27. Lee Kuan Yew, "The Grand Master's Insights on China, the United States and the World" (Belfer Center Studies in International Security, 2013), 20.

28. Steve Case, *The Third Wave: An Entrepreneur's Vision of the Future* (Simon & Schuster, 2016), 146.

29. Julia Powles and Carissa Veliz, "How Europe Is Fighting to Change Tech Companies' 'Wrecking Ball' Ethics," *Guardian*, January 30, 2016.

30. Adam Thomson, Richard Waters, and Vanessa Houlder, "Raid on Google's Paris Office Raises Stakes in Tax Battle with US Tech," *Financial Times*, May 24, 2016.

31. Madhumita Murgia and Duncan Robinson, "Google Faces EU Curbs on How It Tracks Users to Drive Adverts," *Financial Times*, December 13, 2016.

32. Duncan Robinson and David Bond, "Tech Groups Warned on 'Fake News,'" *Financial Times*, January 31, 2017.

33. Ibid.

34. "Terror and the Internet," *Economist*, June 10, 2017.

35. Rachel Stern, "Germany's Plan to Fight Fake News," *Christian Science Monitor*, January 9, 2017.

36. Nick Hopkins, "How Facebook Flouts Holocaust Denial Laws Except When It Fears Being Sued," *Guardian*, May 24, 2017.

37. Melissa Eddy and Mark Scott, "Delete Hate Speech or Pay Up, Germany Tells Social Media Companies," *New York Times*, June 30, 2017.

38. Ben Riley-Smith, "Parliament to Grill Facebook Chiefs over 'Fake News,'" *Telegraph*, January 14, 2017.

39. Robert Tait, "Czech Republic to Fight 'Fake News' with Specialist Unit," *Guardian*, December 28, 2016.

40. Madhumita Murgia and Hannah Kuchler, "Facing Down Fake News," *Financial Times*, May 2, 2017.

41. Seth Fiegerman, "Facebook Adding 3,000 Reviewers to Combat Violent Videos," *CNN*, May 3, 2017.

42. Kelly Couturier, "How Europe Is Going After Google, Amazon and Other U.S. Tech Giants," *New York Times*, April 20, 2016.

43. Sam Schechner and Stu Woo, "EU to Get Tough on Chat Apps in Win for Telecoms," *Wall Street Journal*, September 11, 2011.

44. Mark Scott, "Facebook Ordered to Stop Collecting Data on WhatsApp Users in Germany," *New York Times*, September 27, 2016.

45. Olivia Solon, "How Much Data Did Facebook Have on One Man? 1,200 Pages of Data in 57 Categories," *Wired*, December 28, 2012.

46. Murad Ahmed, Richard Waters, and Duncan Robinson, "Harbouring Doubts," *Financial Times*, October 11, 2015.

47. Luigi Zingales and Guy Rolnik, "A Way to Own Your Social-Media Data," *New York Times*, June 30, 2017.

48. Duncan Robinson, "Web Giants Sign Up to EU Hate Speech Rules," *Financial Times*, May 31, 2016.

49. Glyn Moody, " 'Google Tax' on Snippets Under Serious Consideration by European Commission," *Ars Technica UK*, March 24, 2016.

50. Ahmed, Waters, and Robinson, "Harbouring Doubts."

Chapter Seven

1. Quentin Hardy, "The Web's Creator Looks to Reinvent It," *New York Times*, June 7, 2016.

2. Ben Tarnoff, "New Technology May Soon Resurrect the Sharing Economy in a Very Radical Form," *Guardian*, October 17, 2016.

3. "Does Deutschland Do Digital?" *Economist*, November 21, 2015.

4. Guy Chazan, "Germany's Digital Angst," *Financial Times*, January 26, 2017.

5. Matthew Karnitschnig, "Why Europe's Largest Economy Resists New Industrial Revolution," *Politico*, September 14, 2016.

6. Jonathan Haynes and Alex Hern, "Google to Build Adblocker into Chrome Browser to Tackle Intrusive Ads," *Guardian*, June 2, 2017.

7. Ian Leslie, "Advertisers Trapped in an Age of Online Obfuscation," *Financial Times*, February 28, 2017.

8. Robert Thomson, "Fake News and the Digital Duopoly," *Wall Street Journal*, April 5, 2017.

9. Will Heilpern, "A 'Zombie Army' of Bots Is Going to Steal $7.2 Billion from the Advertising Industry This Year," *Business Insider*, January 20, 2016.

10. John Gapper, "Regulators Are Failing to Block Fraudulent Ads," *Financial Times*, February 3, 2016.

11. Steven Perlberg, "New York Times Readies Ad-Free Digital Subscription Model," *Wall Street Journal*, June 20, 2016.

12. Ken Doctor, "Behind the Times Surge to 2.5 Million Subscribers," *Politico*, December 5, 2016.

13. Gordon E. Moore, "Cramming More Components onto Integrated Circuits," *Electronics*, April 19, 1965.

14. Ralph Nader, *Unsafe at Any Speed: The Designed-In Dangers of American Automobiles* (Simon & Schuster, 1965), vi.

15. Christopher Jensen, "50 Years Ago, 'Unsafe at Any Speed' Shook the Auto World," *New York Times*, November 26, 2015.

16. Ibid.

17. Lee Rainie, "The State of Privacy in Post-Snowden America," Pew Research Center, September 21, 2016.

Chapter Eight

1. Huw Price, *Time's Arrow and Archimedes' Point* (Oxford University Press, 1996), 6.

2. For a lucid introduction to Huw Price's ideas about the block universe theory of time, hear his interview on the podcast show *Philosophy Bites*: "Huw Price on Backward Causation," *PhilosophyBites.com*, July 15, 2012.

3. Price, *Time's Arrow and Archimedes' Point*, 4.

4. Roger Parloff, "AI Partnership Launched by Amazon, Facebook, Google, IBM, and Microsoft," *Fortune*, September 28, 2016.

5. Tad Friend, "Sam Altman's Manifest Destiny," *New Yorker*, October 10, 2016.

6. Rana Foroohar, "Echoes of Wall Street in Silicon Valley's Grip on Money and Power," *Financial Times*, July 3, 2017.

7. Friend, "Sam Altman's Manifest Destiny."

8. Jim Yardley, "An American in a Strange Land," *New York Times Magazine*, November 6, 2016.

9. Ibid., 48–51.

10. Robert Frank, "At Last, Jeff Bezos Offers a Hint of His Philanthropic Plans," *New York Times*, June 15, 2017.

11. John Thornhill, "Zuckerberg and the Politics of Soft Power," *Financial Times*, April 3, 2017.

12. Anjana Ahuja, "Silicon Valley's Largesse Has Unintended Consequences," *Financial Times*, April 26, 2017.

13. David Callahan, *The Givers: Wealth, Power and Philanthropy in a New Gilded Age* (Knopf, 2017), 112–35.

14. Edward Luce, "What Zuckerberg Could Learn from Buffet," *Financial Times*, December 5/6, 2015.

15. Deepa Seetharaman, "Zuckerberg Lays Out Broad Vision for Facebook in 6,000-Word Mission Statement," *Wall Street Journal*, February 16, 2017.

16. Steven Waldman, "What Facebook Owes to Journalism," *New York Times*, February 21, 2017.

17. Olivia Solon, "Priscilla Chan and Mark Zuckerberg Aim to 'Cure, Prevent and Manage' All Disease," *Guardian*, September 22, 2016.

18. Christian Davenport, "An Exclusive Look at Jeff Bezos' Plan to Set Up Amazon-Like Delivery for 'Future Human Settlement' of the Moon," *Washington Post*, March 2, 2017.

19. Mark Harris, "Revealed: Sergey Brin's Secret Plans to Build the World's Biggest Aircraft," *Guardian*, May 26, 2017.

20. Tony Romm, "Mark Pincus and Reid Hoffman Are Launching a New Group to Rethink the Democratic Party," *Recode*, July 3, 2017.

21. Benjamin Mullin, "Craig Newmark Foundation Gives Poynter $1 Million to Fund Chair in Journalism Ethics," *Poynter*, December 12, 2016.

22. Ken Yeung, "Facebook, Mozilla, Craig Newmark, Others Launch $14 Million Fund to Support News Integrity," *Venture Beat*, April 2, 2017.

23. Lisa Veale, "Kapor Center Establishes Oakland as the Epicenter of Tech for Social Justice," *Oaklandlocal.com*, November 6, 2014.

24. Mitch and Freada Kapor, "An Open Letter to the Uber Board and Investors," *Medium*, February 23, 2017.

25. Sarah Lacey, "After McClure Revelations, 500 Startups Lp Mitch Kapor Says He'll Ask for His Money Back," *Pando*, June 30, 2017.

26. Josh Constine, "Uber Investors Who Called It 'Toxic' Are Satisfied by Plans for Change," *Techcrunch*, June 15, 2017.

27. Annie Sciacci, "At New Oakland Hub, Tech Diversity Is Key," *Mercury News*, July 19, 2016.

Chapter Nine

1. Michael S. Malone, "Silicon and the Silver Screen," *Wall Street Journal*, April 16, 2017.

2. Jordan Crucchiola, "Taylor Swift Is the Queen of the Internet," *Wired*, June 22, 2015.

3. Matthew Garrahan, "Pop Stars Complain to Brussels over YouTube," *Guardian*, June 29, 2016.

4. Christopher Zara, "The Most Important Law in Tech Has a Problem," *Backchannel*, January 3, 2017.

5. John Naughton, "How Two Congressmen Created the Internet's Biggest Names," *Guardian*, January 17, 2017.

6. Rob Levine, "Taylor Swift, Paul McCartney Among 180 Artists Signing Petition for Digital Copyright Reform," *Billboard*, June 20, 2016.

7. Debbie Harry, "Music Matters. YouTube Should Pay Musicians Fairly," *Guardian*, April 26, 2016.

8. Rob Davies, "Google Braces for Questions as More Big-Name Firms Pull Adverts," *Guardian*, March 19, 2017.

9. Jessica Gwynn, "AT&T, Other U.S. Advertisers Quit Google, YouTube over Extremist Videos," *USA Today*, March 22, 2017.

10. Joe Mayes and Jeremy Kahn, "Google to Revamp Ad Policies After U.K., Big Brands Boycott," *Bloomberg*, March 17, 2017.

11. Madhumita Murgia, "Google Unveils Advertising Safeguards as Backlash over Extremist Videos Rises," *Financial Times*, March 22, 2017.

12. Jack Marshall and Jack Nicas, "Google to Allow 'Brand Safety' Monitoring by Outside Firms," *Wall Street Journal*, April 3, 2017.

13. Travis M. Andrews, "YouTube Announces Plan 'to Fight Online Terror,' Including Making Incendiary Videos Difficult to Find," *Washington Post*, June 19, 2017.

14. Matthew Garrahan, "Advertisers Skeptical on Google Ad Policy Changes," *Financial Times*, March 22, 2017.

15. Maya Kosoff, "Zuckerberg Hits Back; Don't Blame Facebook for Donald Trump," *Vanity Fair*, November 11, 2016.

16. Elle Hunt, " 'Disputed by Multiple Fact-Checkers': Facebook Rolls Out New Alert to Combat Fake News," *Guardian*, March 21, 2017.

17. Samule Gibbs, "Google to Display Fact-Checking Labels to Show If News Is True or False," *Guardian*, April 17, 2017.

18. Stefan Nicola, "Facebook Buys Full-Page Ads in Germany in Battle with Fake News," *Bloomberg*, April 17, 2017.

19. Ben Sisario, "Defining and Demanding a Musician's Fair Shake in the Internet Age," *New York Times*, September 30, 2013.

20. Nate Rau, "U.S. Music Streaming Sales Reach Historic High," *USA Today*, March 30, 2017.

21. Ben Sisario and Karl Russell, "In Shift to Streaming, Music Business Has Lost Billions," *New York Times*, March 24, 2016.

22. David Lowery, "My Song Got Played on Pandora 1 Million Times and All I Got Was $16.89, Less Than What I Make from a Single T-Shirt Sale!" *Trichordist*, June 24, 2013.

23. Robert Levine, "David Lowery, Cracker Frontman and Artist Advocate, Explains His $150 Million Lawsuit Against Spotify: Q&A," *Billboard*, April 7, 2016.

24. Mark Yarm, "One Cranky Rocker Takes on the Entire Streaming Music Business," *Bloomberg*, August 10, 2016.

25. Levine, "David Lowery, Cracker Frontman and Artist Advocate."

26. Farhad Manjoo, "How the Internet Is Saving Culture, Not Killing It," *New York Times*, March 15, 2017.

27. Ibid.

28. Diana Kapp, "Uber's Worst Nightmare," *San Francisco Magazine*, May 18, 2015.

29. "The Gig Economy's False Promise," *New York Times*, April 17, 2017.

30. Rob Davies and Sarah Butler, "UK Workers Earning £2.50 an Hour Prompts Calls for Government Action," *Guardian*, July 6, 2017.

31. "The Gig Economy's False Promise."

32. Jia Tolentino, "The Gig Economy Celebrates Working Yourself to Death," *Guardian*, March 22, 2017.

33. Chantel McGee, "Only 4 Percent of Uber Drivers Remain on the Platform a Year Later, Says Report," CNBC, April 20, 2017.

34. More, *Utopia*, 82.

35. Hannah Levintova, "Meet 'Sledgehammer Shannon,' the Lawyer Who Is Uber's Worst Nightmare," *Mother Jones*, December 30, 2015.

36. Kapp, "Uber's Worst Nightmare."

37. Barney Jopeson and Leslie Hook, "Warren Lashes Out Against Uber and Lyft," *Financial Times*, May 19, 2016.

38. Ellen Huet, "What Really Killed Homejoy? It Couldn't Hold On to Its Customers," *Forbes*, July 23, 2015.

39. Carmel Deamicis, "Homejoy Shuts Down After Battling Worker Classification Lawsuits," *Recode*, July 17, 2015.

40. Anna Louie Sussman and Josh Zumbrun, "Gig Economy Spreads Broadly," *Wall Street Journal*, March 26–27, 2016.

41. Nick Wingfield, "Start-Up Shies Away from the Gig Economy," *New York Times*, July 12, 2016.

42. Seth D. Harris and Alan B. Krueger, "A Proposal for Modernizing Labor Laws for Twenty-First-Century Work: The 'Independent Worker,'" Hamilton Project, December 2015.

43. Tim Harford, "An Economist's Dreams of a Fairer Gig Economy," *Financial Times*, December 20, 2015.

44. Nick Wingfield and Mike Isaac, "Seattle Will Allow Uber and Lyft Drivers to Form Unions," *New York Times*, December 14, 2015.

45. Chris Johnston, "Uber Drivers Win Key Employment Case," *BBC News*, October 28, 2016.

46. Jessica Floum, "Uber Settles Lawsuit with SF, LA over Driver Background Checks," *San Francisco Chronicle*, April 7, 2016.

47. Rich Jervis, "Austin Voters Reject Uber, Lyft Plan for Self-Regulation," *USA Today*, May 8, 2016.

48. Sam Levin, "Elizabeth Warren Takes on Airbnb, Urging Scrutiny of Large-Scale Renters," *Guardian*, July 13, 2016.

49. Matt Payton, "Berlin Bans Airbnb from Renting Apartments to Tourists in Move to Protect Affordable Housing," *Independent*, May 1, 2016. Also Natasha Lomas, "Airbnb Faces Fresh Crackdown in Barcelona as City Council Asks Residents to Report Illegal Rentals," *Techcrunch*, September 19, 2016.

50. Caroline Davies, "Iceland Plans Airbnb Restrictions amid Tourism Explosion," *Guardian*, May 30, 2016.

51. Rob Davies, "UberEATS Drivers Vow to Take Pay Protest to London Restaurants," *Guardian*, August 26, 2016.

52. Seth Fiegerman, "Uber Drivers to Join Protest for $15 Minimum Wage," *CNN Money*, November 28, 2016.

53. Noam Scheiber and Mike Isaac, "Uber Recognizes New York Drivers' Group, Short of a Union," *New York Times*, May 10, 2016.

54. Emma G. Fitzsimmons, "New York Moves to Require Uber to Provide Tipping Option on Its App," *CNBC*, April 17, 2017.

55. Farhad Manjoo, "One Way to Fix Uber: Think Twice Before Using It," *New York Times*, June 14, 2017.

56. Mike Isaac, "Uber CEO to Leave Trump Advisory Council After Criticism," *New York Times*, February 2, 2017.

57. Josh Lowensohn, "Uber Gutted Carnegie Mellon's Top Robotics Lab to Build Self-Driving Cars," *Verge*, May 19, 2015.

58. Johana Bhuiyan, "Inside Uber's Self-Driving Car Mess," *Recode*, March 24, 2017.

59. More, *Utopia*, 18.

Chapter Ten

1. More, *Utopia*, 51.

2. Ibid., 50.

3. Ibid., 53.

4. John Thornhill and Ralph Atkins, "Money for Nothing," *Financial Times*, May 27, 2016.

5. "Sighing for Paradise to Come," *Economist*, June 4, 2016.

6. Yuval Noah Harari, "The Meaning of Life in a World Without Work," *Guardian*, May 8, 2017.

7. Andy Stern, *Raising the Floor: How a Universal Basic Income Can Renew Our Economy and Rebuild the American Dream* (PublicAffairs, 2016).

8. John Thornhill, "A Universal Basic Income Is an Old Idea with Modern Appeal," *Financial Times*, March 14, 2016.

9. More, *Utopia*, 51.

10. Rutger Bregman, *Utopia for Realists: How We Can Build the Ideal World* (Little Brown, 2017).

11. Martin Ford, *Rise of the Robots: Technology and the Threat of a Jobless Future* (Basic Books, 2016).

12. Albert Wenger, *World After Capital* (worldaftercapital.org 2016).

13. Erik Brynjolfsson and Andrew McAfee, *The Second Machine Age: Work, Progress and Prosperity in a Time of Brilliant Technologies* (Norton, 2014), 213.

14. Lee Rainie, "The Future of Jobs and Jobs Training," Pew Research Center, May 3, 2017.

15. Danielle Pacquette, "Bosses Believe Your Work Skills Will Soon Be Useless," *Washington Post*, May 3, 2017.

16. Ibid.

17. Rainie, "The Future of Jobs and Jobs Training."

18. Matthew Crawford, *Shop Class as Soulcraft: An Inquiry into the Value of Work* (Penguin, 2010).

19. Esther Wojcicki and Lance Izumi, *Moonshots in Education: Launching Blended Learning in the Classroom* (Pacific Research Institute, 2014).

20. Gillian Tett, *The Silo Effect: The Peril of Expertise and the Promise of Breaking Down Barriers* (Simon & Schuster, 2015).

21. Danielle Muoio, "Google and Alphabet's 20 Most Ambitious Moonshot Projects," *Business Insider*, February 13, 2016.

22. Sean Coughlan, "Pisa Tests: Singapore Top in Global Education Rankings," *BBC News*, December 6, 2016.

23. Abby Jackson and Andy Kiercz, "The Latest Ranking of Top Countries in Math, Reading, and Science Is Out—and the US Didn't Crack the Top 10," *Business Insider*, December 6, 2016.

24. Natasha Singer, "How Silicon Valley Pushed Coding into American Classrooms," *New York Times*, June 27, 2017.

25. Natasha Singer, "The Silicon Valley Billionaires Remaking America's Schools," *New York Times*, June 6, 2017.

26. Ibid.

27. Alter, *Irresistible*, 2.

28. Matt Richtel, "A Silicon Valley School That Doesn't Compute," *New York Times*, October 22, 2011.

Conclusion

1. Polanyi, *The Great Transformation*, 3.

2. See, for example: Don Tapscott, *Growing Up Digital: The Rise of the Net Generation* (McGraw-Hill, 1999).

3. David Sax, *The Revenge of Analog: Real Things and Why They Matter* (Public Affairs, 2017).

4. Jordan Passman, "Vinyl Sales Aren't Dead: The 'New' Billion Dollar Music Business," *Forbes*, January 12, 2017.

5. Laura Hazard Owen, "News Apps Are Making a Comeback. More Young Americans Are Paying for News. 2017 Is Weird," NiemanLab, June 21, 2017.

6. Thomas B. Edsall, "The End of the Left and the Right as We Know Them," *New York Times*, June 22, 2017.

7. Elena Cresci, "Meet the 75%: The Young People Who Voted to Remain in the EU," *Guardian*, June 24, 2016.

8. William A. Galston and Glara Hendrickson, "How Millenials Voted This Election," *Brookings.edu*, November 21, 2016.

9. Thornhill, "Zuckerberg and the Politics of Soft Power."

INDEX